FIRE

A BRIEF HISTORY

FIRE

A BRIEF HISTORY

Stephen J. Pyne

Foreword by William Cronon

THE BRITISH MUSEUM PRESS

Also by Stephen J. Pyne

CYCLE OF FIRE

World Fire: The Culture of Fire on Earth

Vestal Fire: An Environmental History, Told through Fire, of Europe and Europe's Encounter with the World

Fire In America: A Cultural History of Wildland and Rural Fire

Burning Bush: A Fire History of Australia

The Ice: A Journey to Antarctica

Cycle of Fire is part of Weyerhaeuser Environmental Books, published by the University of Washington Press under the general editorship of William Cronon

First published in the United States of America
by the University of Washington Press, Seattle

Published in Great Britain in 2001 by The British Museum Press
A division of The British Museum Company Ltd
46 Bloomsbury Street, London WC1B 3QQ

British Library Cataloguing in Publication Data
A catalogue record for this book is available from the British Library

ISBN 0-7141-2762-0

Printed in Canada

To Sonja, Lydia, Molly
who have watched it come full circle

CONTENTS

ACKNOWLEDGMENTS

The grand cycle of fire on Earth: that is the big subject of this small book. It is also, in lesser form, its context, for it is my hope that *Fire: A Brief History* will bring, if not final closure, at least a degree of condensation to the Cycle of Fire suite. In truth, this slim volume can pretend to be little more than a candle to the historical firestorm that it introduces. Probably, too, it would be easier to square a circle than to tweak the Cycle's many bulky narratives into the triangular three-fires conceit of *Fire: A Brief History.* Yet the conceit does have a kind of natural logic. If we can reduce fire to the chemistry of a mere three factors, we should be allowed to do no less for its history.

The publication of the Cycle suite began with discussions about this volume. When William Cronon approached me about contributing to the Weyerhaeuser Environmental Books series, an introductory book on fire was the first project he proposed. We agreed that *Vestal Fire*, which I was rabid to write, would precede the projected *Fire*, while my previous publications on fire would be reprinted over a period of several years. Bill possesses an unmatched blend of intellectual vigor and editorial tolerance. My itch to evoke rather than explain has exasperated him more than once; and with regard to this long-delayed volume, he has shown almost preternatural patience. This time I have tried to emulate his passion for clarity and his empathy for the taught audience, although I cannot hope to equal him on such matters. He has my deep gratitude.

As do Julidta Tarver and the staff of the University of Washington Press. At times our correspondence has piled to the point of seeming self-detonation, yet Julidta remained ever pointed, pragmatic, and unfailingly cheerful. No writer could ask for a better editor.

The list of long-sufferers, however, must begin and end with my family. More than once they have asked when this project might conclude. The answer is clear: it won't. But the greatest of the cycles it tells is the one we have collectively lived. Lydia, our first-born, arrived while I was at the National Humanities Center writing *Fire in America*; she has now helped me edit *Fire: A Brief History*, while Molly is ready for a child of her own. The hearth fire has never burned brighter.

FOREWORD:
SMALL BOOK, BIG STORY

William Cronon

For me, one of the chief pleasures and privileges of serving as general editor of Weyerhaeuser Environmental Books has been the opportunity to introduce each volume in the series with a brief essay that shares with readers my own enthusiasm for the work of its author. Because Stephen Pyne is the most prolific of our writers, I've had the chance to introduce no fewer than six of the books in which he has taken on the daunting task of narrating the entire human history of fire on Earth. I cannot claim to have anything like Steve's depth of knowledge or consuming passion for the subject of fire, so I've occasionally worried whether I might eventually run out of things to say in reflecting on the scholarly achievement these works represent. But quite the opposite is true of the small book you now hold in your hands. Although it is the slenderest volume we have published in Steve's Cycle of Fire series, it is also among the most remarkable. Indeed, I feel a special pride of vicarious authorship about it, for reasons I'd like to explain in this foreword.

When I first learned that I would be editing the Weyerhaeuser series, I began casting about for books and authors that might be ideally suited to publish in it. Steve Pyne was among the very first who occurred to me, and I therefore approached him to see whether he might be interested in writing a book which, frankly and selfishly, I had long wanted to read and which only he could write. My reasons for this had much to do with the qualities that make Steve such an unusual figure in the field of environmental history. His defining virtue as a scholar is of course his extraordinarily encyclopedic knowledge of fire and the many ways that human beings have interacted with it since our hominid ancestors first discovered the trick of capturing its lightninglike magic and turning it toward their own ends. No one has ever known or cared more about this subject, surely, than Steve Pyne, and he has made a lifework of sharing what he knows in print. His subject is so vast that he has leapt around the world, century by century and continent by continent, in pursuit of his quarry, filling literally thousands of pages in the process. So expansive is his oeuvre, and so intricate the fine-grained textures and patterns he reveals to readers, that I long ago began to suspect that many of those readers might welcome a road map to help them navigate the vast intellectual landscape that Steve lays out before them.

This book is that road map.

The idea for *Fire: A Brief History* as I first proposed it to Steve was something very different from the five monumental volumes, collectively entitled Cycle of Fire, in which he has devoted hundreds of pages each to the fire histories of the United States, Eurasia, Australia, and elsewhere. In those magisterial books, God and the Devil are both in the details, so that keeping track of trees and forest together is both the challenge and the reward that readers face. Cycle of Fire is, in effect, an enormous intellectual mansion with many, many rooms, covering so much territory under one roof that inexperienced readers can be forgiven for occasionally losing their bearings while wandering its corridors—not because its author is a confusing guide, but because his subject is so demanding, and so unfamiliar to most of us. I suggested to Steve that readers might find their way more easily through his other books if they had on hand a much shorter volume offering a bird's-eye view of the whole. What readers needed, to combine the metaphors, was a blueprint of the mansion, and a way to survey the surrounding countryside by stepping back from the individual trees so as to grasp the shape of the immense forest that contains them.

The result is this little book. Although it is certainly the shortest volume in the Cycle of Fire sequence, it is also, arguably, the most ambitious. Never before has Steve Pyne narrated the entire story of earthly fire in so few pages. Never before has he sought to distill his scholarly insights into a handful of core defining observations. Never before have the intricacies of fire history—world-wide and through the whole sweep of human history—been revealed in such stunning relief. The book is a triumph not of abridgement but of concentrated distillation. What I have said before about Steve's other books is even more true of this one: for those willing to gaze through the unusual lens it offers our eyes, it can change the way we see and understand the world.

At the center of *Fire: A Brief History* is an unfolding narrative structure that divides fire's human history into a series of overlapping epic chapters. In the beginning was nature's flame, the almost irresistible chemical tendency toward oxidation that has defined all life on Earth since the moment photosynthesizing plants began to produce the profoundly unstable oxygen-rich atmosphere that has ever since been among the most defining features of this peculiar planet. When early hominids learned to carry and build this fire at will, making it their own, they began the long process whereby human beings have transformed the Earth by redirecting the complex routes that flames have burned across it.

The next chapter in this process of fire's coevolution with humanity was the invention of agriculture and the very different fire dynamics it necessarily entails: fire to clear fields, fire to change the composition of wild and domesticated vegetation, fire alternately bound and released in cycles that sometimes seemed increasingly under human control, and sometimes, devastatingly, not. The fires that have burned under this second, agricultural regime have brought a complex remapping of the Earth's surface, extending fire's reach in some regions and habitats while suppressing it in others. The consequences of this human manipulation of terrestrial fire ecology have been so subtle and profound that we are only now beginning to understand them.

Finally, we come to industrial fire, which has been increasingly dominant across ever greater portions of the Earth's surface for the past three centuries, even as natural and agricultural fires have persisted alongside it. Industrial fire has been characterized by two tendencies: it burns in carefully controlled spaces from which energy and motive power can be extracted, and its source is drawn not from the immediate flux of calories emanating from the sun but rather from buried fossil fuels that make it possible for sunlight hundreds of millions of years old to shine on Earth once again. Industrial fire has tremendously increased the human power to manipulate the planet for good or ill, augmenting to an astonishing degree our powers of production while at the same time giving us terrifying new tools for rendering into dust that which we wish to destroy. This is the era in which we now live, whose ending we cannot know but whose fate we cannot help but share. What we *can* know is that the fate of humanity, like the fate of the Earth, is tied to the fires that have made the world as we know it—the fires whose history is told as well in this book as it has ever been told before.

Steve Pyne is a historian, not a prophet, and this small book cannot solve the riddle of fire's future: it cannot predict what forms of fuel might avert future energy crises any more than it can predict what forms of burning might avert future global warming. What it can do is help explain why things like energy crises and global warming—to say nothing of rural and urban land use, human food supplies, forest management, industrial production, and the ever-present threat of wildfire—are bound not just to the history of humanity but to the history of fire as well. Indeed, the great insight of this book is that the two are so inextricably bound to one another that it finally makes no sense to tell their stories separately. No other book in Steve Pyne's Cycle of Fire has made this point more persuasively. The fire that has burned in humanity's hearth from the beginning, the fire with which we have remade the world, is a profoundly double-edged symbol both of our Promethean power to control the Earth . . . and of the frustratingly unexpected limits we repeatedly encounter in our exercise of that power. If one wants to understand just how completely the story of the human past is also the story of fire on Earth, there is no better place to start than this small book.

INTRODUCTION: KINDLING

There was a time when the Earth did not burn; when oxygen did not soak its atmosphere, when plants did not encrust its lands. But for more than 400 million years the planet has burned. In some places and times, fire has trimmed and pruned flora; in others, it has hewn whole biotas; for virtually all it has simply been there like floods and earthquakes, like the winds, droughts, seasons, browsers, and lightning with which it is associated. For almost all the span of terrestrial life, fire has continued, to varying degrees, as an environmental presence, an ecological process, and an evolutionary force. Fuel, oxygen, heat—that is fire's triangle. At various times the play of fire's triangle has been cyclic, singular, evolutionary, but once created it has always endured.

Even on a planet as distinctive as ours, fire's story is special. Fire is unique to Earth and our seizure of it unique to humanity. Although space exploration has revealed that other planets hold some of the components for combustion, none have all of them or the context by which to mingle fuels, oxygen, and spark into the explosive reaction we call fire. So, too, while all species modify the places in which they live and many can modify fire's environment, only humans can, within limits, start and stop fire at will. Other organisms can trash forests, uproot shrubs, denude grasses, promote seedlings, choose one plant rather than another. Some organisms breed in fresh charcoal, some forage among ash and hunt along flame's flank and through clouds of smoke, some self-immolate with a vigor that bestows upon them a selective advantage in comparison with less fire-prone neighbors, some like Philippine tarsiers may even grasp embers in their claws or like Australian kites seize the embers in their talons and redeposit them elsewhere, probably by accident, perhaps by intent. Nicotine-addicted chimps will toy with burning cigarettes. But only humans can kindle fire, sustain it, and spread it beyond its natural habitats. Only humanity has become, for the biosphere, the keeper of the vital flame. Fire's story is a story of the Earth and, as myths emphatically insist, a story of ourselves.

The narrative for fire has an intrinsic logic. The first movement involves the creation of combustion, a reaction which, simply put, takes apart what photosynthesis brings together. With an atmosphere fluffed by oxygen and lands lathered in plants, combustion could leave cells and burn where wind and fuel could take it. At that point one can speak of fire. The earliest charcoal preserved in the geologic record dates back to the Devonian.

But fire is a catalyst, it takes on the character of its context, it synthesizes its surroundings. The fires of the Paleozoic were undoubtedly different from those of recent times. Probably much of the Earth lacked fire altogether and other parts had it in spasms. Certainly immense stocks of biomass failed to burn and were simply buried. Parts of the Earth continue to combust from strictly natural causes, though little of that burning now occurs in completely natural ways.

All this changed profoundly when early hominids captured fire and then devised ways to kindle it on demand. Fire became a species monopoly: it flourished as a unique power humans would never willingly share with other creatures. But, again, fire can burn only what its surroundings furnish. Some landscapes could be burned easily, some not at all. Anthropogenic fire could thrive only where nature fed it. This left large chunks of Earth unburned, and other chunks that burned according to different regimes.

To leverage their fire power further, humans needed to manipulate fuel as they did ignition. From the perspective of fire history, this is the meaning of agriculture. Fuel could be created by slashing or browsing, grown by planting and fallowing, burned according to the rhythms of field and pasture. The dominion of fire expanded enormously. Only the most formidable lands remained outside its reach. The greatest extent of open flame resulted from the far-flung colonizings of agriculture, most of which involved some rotation by which fuels were fashioned and then burned.

Still, human fire power was only as great as the stocks of fuel that nature, with human contrivance, could be made to provide. Serious limits remained: only so much biomass could result from cutting, planting, and fallowing. These barriers fell when, outfitted with combusting machines, people reached into the geologic past and exhumed fossil biomass. For fire history, this marks the moment of industrialization. The limitations on fire reside no longer in its sources—ignition and fuel—

but in the sinks such as the atmosphere that must receive combustion's unbounded byproducts.

All three fires thrive today. How industrial combustion plays against natural fire and the variants of anthropogenic fire is, in particular, the unsettled story of fire's current geography. While the three groupings of fire compete, each with the others, they also coexist. What endures is fire in one form or another. What endures, too, is the unique status of humanity as the keeper of those flames. Fire tracks, as perhaps no other index can, the awesome, stumbling, unexpected, implacable, fascinating course of humanity's ecological agency. The story of one cannot be told except through the other.

A BRIEF HISTORY

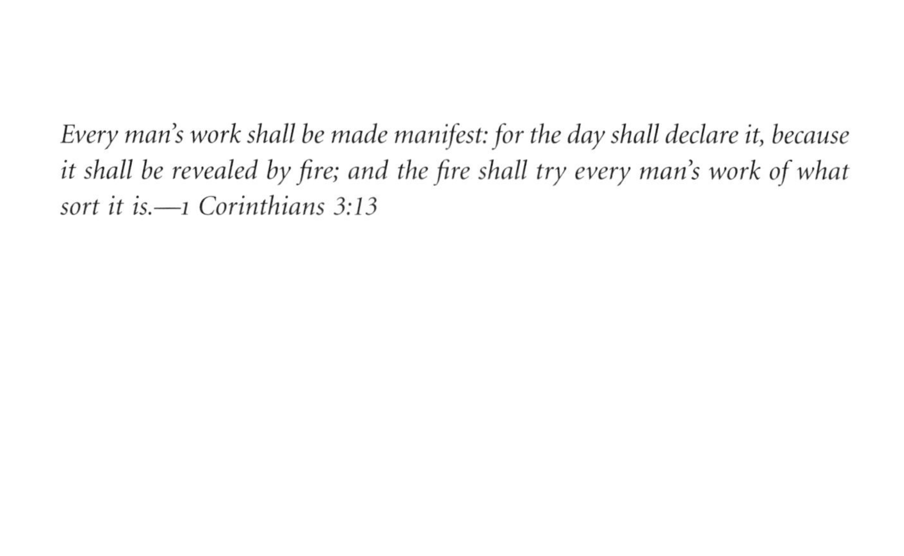

Every man's work shall be made manifest: for the day shall declare it, because it shall be revealed by fire; and the fire shall try every man's work of what sort it is.—1 *Corinthians 3:13*

Chapter One

Fire and Earth

CREATING COMBUSTION

According to many myths, we became truly human only when we acquired fire. So it is natural to assume a parallel awakening for the place we live. Rather, the Earth likely simmered through more than four billion years before its biotic broth boiled over. Some of fire's components the ancient Earth acquired only after long eons. Even more critically, those ingredients needed a durable context in which to mingle. The parts had to combine and do so consistently. Combustion has its creation story. Fire has its history.

Of fire's three essential elements, only the heat of ignition thrived on the early Earth. Oxygen did not begin to collect until the last two billion years, and did not begin to approach modern quantities until roughly 500 million years ago. Land plants suitable to carry combustion did not become abundant until 400 million years ago. Before that time the Earth lacked the means to burn regularly or vigorously. It is possible that aquatic biomass might have burned, if a lagoon or marsh dried or storms hurled kelp or algal mats into deep berms where they dried, met lightning or lava, and combusted. But such burns, if they occurred, would little resemble modern fires, and are ecological freaks, never absorbed or ordered within a biological community. Earth's original fires—its colonizing fires—demanded land plants. Probably these consisted of primordial moors, a matrix of near-shore organic peat and reeds. Fires probably first flickered during the early Devonian, roughly 400 million years ago. The most ancient fossil charcoal dates from that epoch.

Since then, fire's evolution has been unending if uneven. Each of combustion's components has existed more or less distinctly from the others, colliding from time to time with a fizz of oxidation or a brilliant burst of burning. But combustion could survive only if it had a consistent and durable context. Over time, fire became itself a synthesizing process, a kind of biochemical flywheel that has helped to balance its separate parts into a coherent whole. It has affected the chemistry of the atmosphere. It has influenced, perhaps profoundly, the character of life. Progressively, the biosphere has absorbed fire and tweaked it to fit a system of biological checks and balances. This was easiest with oxygen and fuel, both the products of life. The absorption of ignition proved more vexing, and had to await the arrival of creatures who could make sparks and heat as easily as they could drill bone and chip flint. Those creatures, of course, were ourselves.

How Fire Came to Be

Casting Sparks

Combustion requires a spark. It needs a jolt of energy to unpack photosynthesized matter, to set off a chain reaction that can release enough surplus energy through oxidation to continue. The early Earth offered several sources: falling rocks, volcanic discharges, extraterrestrial impacts, and lightning. After dead biomass collected in heaps, spontaneous combustion was selectively possible, and after fossil fuels were exposed, coalfields, petroleum seeps, and oil shales could take fire and hold it for centuries, even millennia. But of this geophysical exotica, only lightning is sufficiently consistent and universal to account for the natural history of fire.

Volcanoes are a faux fire, but they have the capacity to kindle real ones. Flowing lava instantly burns what it touches; eruptions often spawn

---

FIGURE 1. The wet and the dry. Patterns of wetting and drying shape ignition as well as fuels. Areas without lightning lack natural fires because there is no spark; but spark alone is not sufficient if heavy rain accompanies it. The geography of lightning is not identical to the geography of lightning-caused fire.

Consider the United States. A map of thunderstorm days (top) shows a concentration of lightning in the Southeast. A map of lightning-kindled forest fires (bottom), however, highlights the West. In particular, the Southwest boasts an ideal formula for fire. A long droughty spring ends in a "monsoon" announced by sporadic summer thunderstorms, beginning in early July. In their first surge, the storms are scattered, some towering over superheated deserts or more commonly over isolated mountains and high rims. Often these thunderstorms are so dry that the rain evaporates before it reaches the ground or soon afterward. There is enough moisture to power a storm but not enough to saturate the surface fuels. As the rainy season progresses, more fires start but fewer become large as the grasses green up and the woody stems flush. (Sources: Schroeder and Buck 1970, and *Yearbook of Agriculture* 1941, both redrawn by the University of Wisconsin Cartographic Lab)

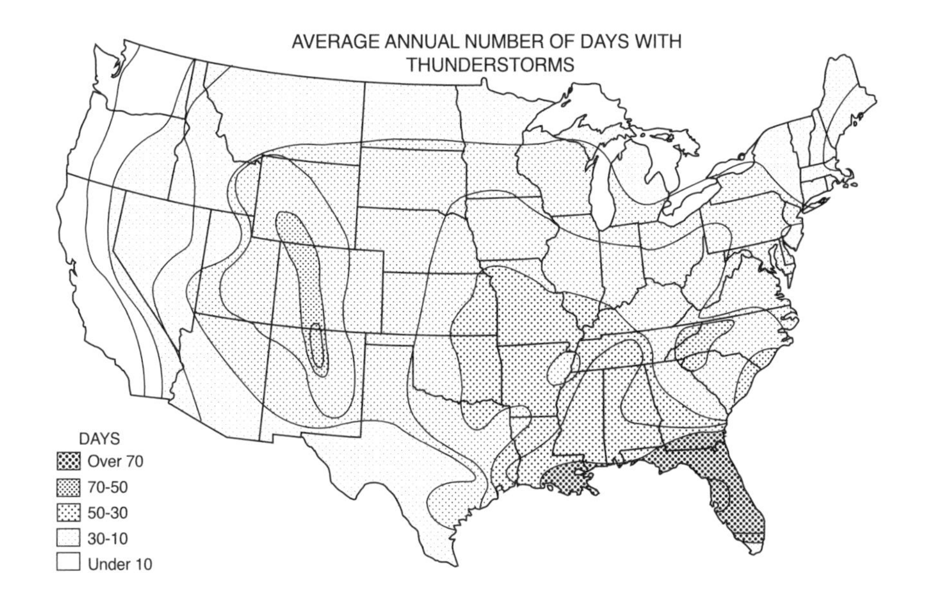
AVERAGE ANNUAL NUMBER OF DAYS WITH THUNDERSTORMS
DAYS
Over 70
70-50
50-30
30-10
Under 10

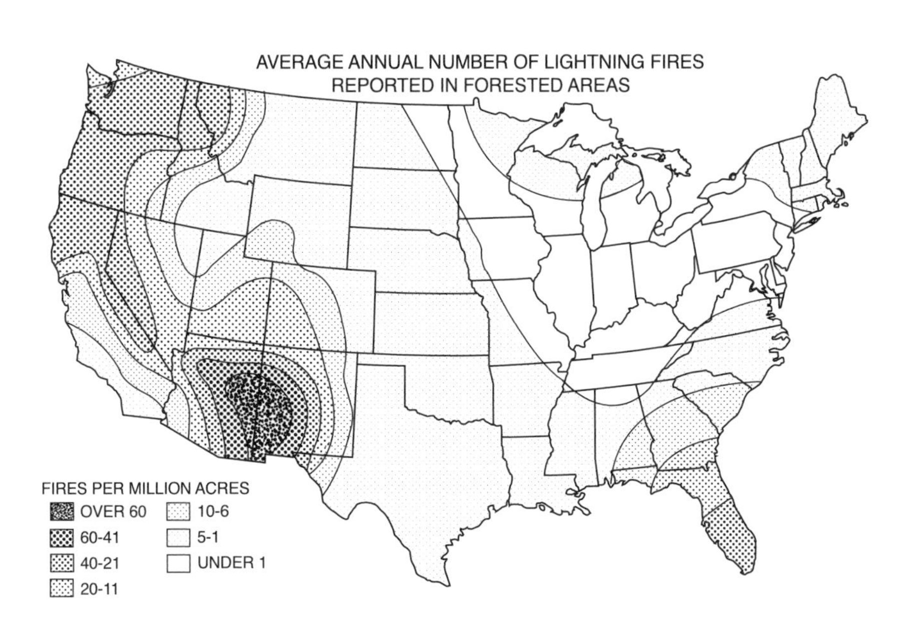
AVERAGE ANNUAL NUMBER OF LIGHTNING FIRES REPORTED IN FORESTED AREAS
FIRES PER MILLION ACRES
OVER 60
60-41
40-21
20-11
10-6
5-1
UNDER 1

thunderstorms, which hurl down lightning like volcanic bombs. But while widespread across geologic time, volcanoes are fixed in geographic space. Moreover, volcanic-kindled fires burn locally; lava or ash covers the burns, as often as not; and one way or another the overall disturbance of the volcano smothers the effects of the fire. In brief, volcanoes are too few, too small, too rare to account for the near-universal realm of fire. Most of the burning Earth is far removed from spark-casting volcanoes.

That leaves lightning. Not every place experiences lightning, and not every lightning-blasted place knows fire. The natural history of lightning fire is lumpy: the fires come in bursts, they crowd in time, they bunch in space. Some years have many, some have few. Although some places never know them, some feel them annually, or until climate reshuffles the deck of places wetted and dried. But its longevity, geography, and concentrated heat mean that lightning clearly accounts for the fact that fire is geologically old and geographically extensive.

Even so, only a tiny fraction of lightning kindles fire. Only one bolt in four reaches the ground. Most of those strike rock or sea, or slam into fuels too wet or shattered or misarranged to burn. Of those that hit something combustible, only one in five has the right properties to convert electrical charge into combustion, the "hot" lightning with high amperage and low voltage. (High-voltage "cold" lightning tends to blast without burning.) Moreover, the storm that looses lightning also dumps rain. What the first can start, the second can stop. The geography of lightning thus overlaps only lightly with the geography of fire. Rather, fire burns along the margins—with the first storms after a long drought, or from dry storms whose veils of rain evaporate before reaching the ground, or in regions prone to severe swings of wet and dry conditions. However often lightning rolls the dice, the house odds remain against fire.

Yet ignition occurs often enough to render lightning fire the vestal flame of the ancient Earth. In some landscapes it is fickle if powerful, rather like hurricanes. In America, for example, dry thunderstorms can charge whole regions with fire. Between 1946 and 1973, "fire busts" in the Northern Rockies splattered the national forests with more than 100 fires a day on 25 occasions; five times, the total exceeded 200 fires. On July 17, 1940, there were 335 fires. Over a ten-day period this same storm kindled 1,488 fires. Between 1960 and 1974, in the national forests of the Southwest five times lightning kindled in excess of 500 fires over a ten-day period. The region averages roughly 2,000 such fires a year. Here lightning fire is as much a chronic presence as storm and sunlight. But wherever it occurs the biota adjusts. Some trees are struck more often than others; curiously, these tend to be species best adapted to survive fire. The lightning bolts that relentlessly restore electrical equilibrium to the Earth also maintain its biological equilibrium.

Yet they do so in peculiar ways. While trees may adapt to lightning, lightning does not adapt to trees. It knows no biological feedback. If lightning has, over geologic time, been the most persistent of fire's elements, it is also the most inflexible. It obeys a geophysical logic, a cold spark without biological control. It matters not to lightning whether it strikes granite or lodgepole pine, a lake or a barn, a sodden log or a snag as parched as kiln-dried lumber. Lightning rips through Jupiter's atmosphere as much as Earth's.

Oxygen and biomass could not ignore the biosphere: life created them. They would have to interact, and combustion would coevolve along with them. But fire's primordial instrument of ignition could exist on its own, leaving fire without a biological means for regulating spark as it did fuel and oxygen. Or rather, it lacked such means until hominids wrested ignition away from lightning's virtual monopoly. From that moment on, the most rigid element of fire's combustion triangle became the most pliant; and a process that had depended on an electrical charge—its bolts as precise as a rifle shot and as random as storm winds—became a global spark as common as grass and as universal as humanity's restless hand and roving mind.

Making Air

Lightning can spark a reaction, but it cannot sustain one. For the act of kindling to yield to self-sustaining fire, free oxygen has to flow into the combustion zone. Yet only in the last two billion years has the Earth succeeded in filling its atmosphere with oxygen on any scale. For several hundred million years thereafter, the atmosphere's oxygen content waxed and waned. During the Carboniferous and Permian, it swelled to perhaps 35 percent, which made possible a general giantism—beetles the size of puppies and dragonflies as big as ravens. By 150 million years ago it stabilized at 21 percent. For this immense shift, without which fire could not exist, the evolution of life is responsible. Plants pumped out more oxygen than the early Earth could absorb.

That early Earth produced some oxygen by splitting volcanically outgassed molecules of water and carbon dioxide. Such photolysis, along with the chemical weathering of metallic oxides, spewed out packets of free oxygen. But the early Earth's atmosphere was a reducing one. Whatever oxygen that photolysis could produce was quickly absorbed as the freed molecules bonded to carbon, hydrogen, iron, and sulfur—all of them ravenous scavengers of free oxygen. The earliest life (perhaps around 3.5 billion years ago) emerged without oxygen. The first photosynthesizers (roughly 3.1 billion years ago) were also anaerobic. The chemical avidity of free oxygen probably threatened them and likely proved toxic.

The shift from an atmosphere empty of oxygen to one rich in it occurred when oxygen sources increased and oxygen sinks diminished. A critical moment came about 2.3 billion years ago when the earliest photosynthesizing prokaryotes appeared in the form of seaborne blue-green algae. They bolstered oxygen's sources on a large scale by releasing it as a byproduct; yet what algae pumped out, rocks soaked up. Vast quantities of iron, sulfur, and especially carbon were oxidizing and settling into the sedimentary lakes of geologic time. By 1.8 billion years ago, oxidized iron had become abundant in the geologic record, followed by carbonates. In the beginning, carbon dioxide had dominated the Earth's atmosphere; but after eons of lithic burial in forms like limestone, it was becoming a mere trace element, replaced in bulk by the more inert nitrogen. Eventually oxygen's sinks began reaching full capacity and free oxygen flooded the atmosphere. The living world, like the geologic, had to accommodate it. Gradually, organisms transformed a threat into an opportunity. Around 1.3 billion years ago, aerobic photosynthesis emerged, further soaking the Earth's air. About 600 million years ago, select organisms learned to exploit the oxygen that surrounded them to split apart what photosynthesis had joined. Aerobic respiration became common, and a chemical poison evolved into a biochemical necessity.

The chemistry of respiration is a chemistry of combustion. When photosynthesized hydrocarbons are jarred by the right shock, they break apart into carbon dioxide, water, and released energy—a kind of "slow" combustion. In brief, outfitted with special enzymes and antioxidants, organisms so accommodated an oxidizing atmosphere that they neutralized a potentially ruinous reaction and then absorbed and redirected it to their own ends. That, by analogy, is what terrestrial life also did when it found itself steeped in oxygen and blasted by lightning—a process of "fast" combustion we call fire.

There is more. Just as fuels exist within a larger biological context, so oxygen exists within an atmosphere. Fire responds to the air mass as a whole as well as to its selective parts. Of course without oxygen the atmosphere would be fire neutral or even a retardant. But other properties of an air mass can shape how a fire behaves. How does oxygen enter the combustion zone? Can flames expand freely upward? Will they spread in one direction rather than another? What fuels are dry? The larger properties of the air mass—its layering and stability, its winds, and its moisture (as relative humidity or storm-dropped precipitation)—will help decide. The structure of air is as vital as its chemistry, and the history of climate as relevant as the history of how the atmosphere evolved.

The question arises then whether fire has influenced the atmosphere within which it burns. Is free-burning fire a vital process in the global oxygen cycle, or simply a geochemical afterthought? Since fire and life have coevolved, have fire and the Earth's atmosphere as well? Surely combustion mediates the exchange of gases between plants and the atmosphere. It frees carbon from plants and it buries carbon as charcoal. But how much? And by regulating carbon, has fire also regulated oxygen?

Probably, but not significantly. Fire accounts for only a small fraction of global respiration. More tellingly, the linkage between oxygen's partial pressure and free-burning fire can be indirect. It is not clear that the great oxygen bubble of the later Paleozoic supported giant fires as it did giant mosquitoes. Sedimentary rocks from that era hold large stocks of charcoal, but even larger reserves of unburned biomass. Higher oxygen content does not directly translate into more fire. Spark and oxygen must still act on organic matter.

In nature, what are most important are the overall characteristics of the surrounding air and those of the fuels. The size and shape of individual particles, their chemistry, their compactness and arrangement in ways that allow oxygen to flow over their surfaces, and above all their moisture content determine whether a fire spreads from one kindled particle to another, whether combustion flames or smolders, and whether fuel burns as a surface flash or a deep-burrowing glow. Oxygen content would have to rise significantly for large, wet boles to burn, and it would have to drop hugely to prevent ignition in small, dry grasses. By whatever feedback fire shapes the atmosphere, it seems to do so through fuels. After all, the photosynthesizing plants that pump oxygen into the air are the same ones that stoke free-burning fire.

Evolving Fuels

Once stabilized, atmospheric oxygen has remained relatively uniform, much more so than ignition. Throughout the known history of fire, oxygen has persisted, a combustion constant. No surface fire has self-extinguished due to its absence. That fact has made fuels the prime biotic controller of fire. The history of combustibles, however, is nothing less than the evolution of terrestrial life.

Life had first to send its spores and extend its tendrils to land. That move exposed its photosynthesized hydrocarbons to oxygen and spark, and removed them, at least fitfully, from the smothering and cooling baths of water. Life's surge onto land injected burnable biomass into an otherwise empty combustion chamber. About the same time that vascular plants began seriously colonizing the Earth, the first evidence of fire appears in the geologic record. Yet if fire could not exist without fuel, neither would fuels—the planet's vegetative cover—exist without the evolutionary and ecological presence of fire. Each has directly shaped the other.

Follow the fuels. A field guide to fire would be a thesaurus of fuel types. Fire has acquired the vigor, subtlety, and endless variety of the organic world. The biochemistry of metabolism determines the chemistry of combustion; the ecology of biotas establishes the ecology of fire; and the evolution of new organisms shapes the evolution of fire regimes. But the reverse has also been true. Over and again fire has synthesized fuel, oxygen, and spark. Those species that could not accept this fact, like those that could not accommodate oxygen and retired to anaerobic niches, were doomed to occupy the apyric environments of the Earth.

Yet many place and periods probably did escape fire. Fire is an event, not a principle. It occurs specifically, not generically. It is easy to imagine large chunks of the Earth or blocks of Earth history that might have evaded the fusion of lightning and hydrocarbons. It is not enough that fuels bulk large on a landscape. There must be kindling—fine fuels, without which lightning blows apart rather than ignites and flame expires rather than renews. The fuels must be dry or else the heat of ignition will be wasted in boiling off the held water. They must be organized in such a way that the combustion zone can spread. Fuels—even dead fuels—are not really dead: they still flourish within a complex biological system. Their availability depends on competition with decomposers and browsers and rival species; on the climatic choreography of sun and wind, drought and rain; and on the crude timing of lightning with seasonal and secular cycles that green up and cure surface plants. It would be easy for a particular place to miss fire, and is strikingly clear from the geologic record that many did.

Locally, yes; everywhere, no. Fire demanded only certain chemical conditions, and whenever they met, it could spring into being. It could, for a while, ecologically expire, then revive. It could vanish for perhaps long eons, then return. Slowly, however, its critical parts began to interact in ways that rendered fire less random, that made its appearance and absence less like the roll of a roulette wheel and more like the give-and-take of prey and predator. It mattered not to lightning if fire happened, and most likely it mattered only marginally to oxygen. But terrestrial life would evolve with fire. Fuels were alive and could influence the character of fire in ways the pure chemistry of oxygen and the pure physics of lightning could not. By means both coarse and delicate, fire could shape the world that shaped it.

A Prehistory for Fire

Sometime between 450 and 400 million years ago, the pieces snapped together with enough force to burn and keep burning. Before that moment, fire did not exist; afterwards, it became almost impossible for it not to. The eccentric ecology of fire has since evolved along with the often lurching evolution of its parts. While the raw chemistry of combustion has remained more or less constant, fire has no more abided unchanged than has climate or life. First Fire's behavior and habitats likely looked different from today's. Triassic fires were probably as distinct as Triassic fauna and the flora they browsed and shaped. Fire's regimes during the Carboniferous, lacking grass, little resembled those typical of the Holocene, loaded with grasslands and grazers.

What were ancient fires like? Simply put, they were like the fuels on which they fed, which makes a reconstruction all the more difficult because so little is known about the range of ancient plants and how they covered the primordial Earth. The mystery is worse for the fine fuels. Small particles of combustibles—pine needles, grasses, small twigs—respond to heat and moisture more quickly than large ones do. They dry and wet faster, ignite and flare more readily. For a propagating fire, logs and peat are combustion sinks rather than sources; they may burn for a long time and release masses of carbon byproducts, but the flaming front rushes along with the small and the quick. Drop a handful of dry needles into a campfire and it will flash. Drop a thick, leafy branch and it will smolder and may go out. So, too, a spreading fire surges and sags with the tempo of the fine fuels. But what were the fine fuels during the Pennsylvanian, or the Jurassic? Long-needled gymnosperms did not evolve until the late Paleozoic, deciduous angiosperms until the late Cretaceous, and grasses not until the Miocene. The pine needles, oak leaves, and bunch grass that carry fire today did not exist.

What then supported fire? If the preserved record is a guide—and it favors the big and the tough rather than the tiny and the volatile—early fires burned amid reedlike psilophytes and pteridophytes, within once-sodden swamps of rotting debris, biomes later enriched with horsetails, woody and soft-leaved ferns, lepidodendroid trees, proto-gymnosperms, and by the Carboniferous fanlike ferns, towering lycopods, and *Calamites* trees with branches that whorled like a maypole. All probably could combust under the proper conditions, and some perhaps could sustain spreading flame. Analogues that exist today burn nicely: palms shrug off fires like raindrops, bracken fern carries flame with the wind, and swamps, drained by drought, readily refill with smoldering ash. Yet, although they undoubtedly combusted, such ancient biotas probably bear no more relation to recognizable wildfires of today than do lepidodendroids to lodgepole pine, or psilophytes to tallgrass prairie.

Did combustion propagate or burn in favorable pockets? Did it reach the crowns of the taller woody plants? Did it flame or smolder along the surface? How often did it return, how frequently did it invade swamps, what kind of chemical cocktail did its smoke transport to the sky? No one knows. But the complexities are even greater, for adaptations to fire are rarely singular. Each trait typically supports several needs. Other species compete with fire for biomass, and what they don't consume they can reshape. Heavy grazing can redistribute or even ruin the surface fuels and halt a spreading fire. Browsers can force plants to elevate their crowns, lofting sensitive tissue away from flames. Plants, animals, and fire quickly make for an ecological three-body problem that is most likely insoluble in its details. To what extent was fire a selective force in evolution? What regimens of fire might have existed? Which components carried fire? All are, at present, unanswerable except by analogy.

The paleontology of fire is a vastly inexact study, and fire's prehistory an act of informed speculation. Reconstructing the dynamics of paleofires from the mineral char of *Pteridium* is like reconstructing the physiology of a dinosaur from a preserved femur, or more to the point, like imagining the biota within which that boned dinosaur lived. Were such fires cold-blooded or hot? Did they normally flame or smolder? Did fire regimes consist of ground fires or crown fires? Jurassic fires may resemble modern ones as much, or as little, as a pteranodon does a condor.

Yet fires there were. Fusain (fossil charcoal) exists as their geologic record. Charcoal is nearly immune to further decomposition: it not only preserves fire, fire preserves it. For the Pennsylvanian period, fusain or semifusain comprises 2 to 13 percent of preserved coal, a number that seems to decline over the era. Yet the fire cycle was, by today's standards, out of balance. Fuel sources far exceeded fire sinks; producers raced ahead of decomposers; and fuels piled up faster than fires could remove them. During the Mesozoic, conifers as well as ferns burned. Both have left charcoal residues that preserve the structure of leaves and woody cells. Such fossils testify to the combustion of both dried and green wood. In the marine deposits of the North Sea, charcoal is frequently "the most common form of fossil plant preservation." But there is far less residue than in previous eras. The epoch ended with a bang, and perhaps a burn. The famous boundary between the Cretaceous and the Tertiary eras—a time of mass extinctions, one of the most vivid breaks in the geologic record—was also evidently a time of mass combustion. Atop the K/T boundary's meteoritic-spawned layer of iridium rests a zone of charcoal that could only have resulted from sustained burning, almost certainly continued long after the geophysical tremors of the impact had passed, probably gorging on the mass-killed woods left as biotic berm. By the Tertiary period, the fusain fraction has fallen to less than 1 percent.*

Fire's historical record is thus wildly uneven. The Earth's combustion economy had no invisible hand that balanced fuel and flame, that assured an equilibrium of constant combustion. The market for fire boomed and crashed. The Carboniferous and Mesozoic were times of saving; the present, a time of spending. Despite the heaps of fusain it deposited, the Carboniferous piled even greater stocks of unburned fuel. Why burning apparently diminished during the Mesozoic is unclear. Fluctuations in oxygen, large-rhythm climatic changes, rearranged continents, a quickening of lignin-rich (and combustion-poor) plants, a proliferation of browers and of browser-hungry carnivores, the flukes of the preserved

*Jennifer M. Robinson, William G. Chaloner, and Timothy P. Jones, "Pre-Quaternary Records of Wildfire," in James S. Clark et al., *Sediment Records of Biomass Burning and Global Change* (Berlin: Springer-Verlag, 1997), p. 264.

record itself—the options are many. Still, a general trend is apparent. As the Earth evolved, the great fire imbalances of the early epochs tended to dampen. The coupling between biomass produced and biomass consumed sharpened. Fire apparently became both more common and less eruptive.

What is clear, however, is that a biologic chasm has existed between what might burn and what did burn. Perhaps the biomass was simply unavailable—matter, not fuel—because the climate lacked the proper wet-dry rhythms to crack open and dry out the vegetation, or because the right animals did not exist to munch and crunch biomes into burnable states, or because so much lay in swampy environs beyond oxidizing (and fire). A no less likely explanation is that ignition was too random and fire too geograpically specialized. Fuels could hide in wet nooks and seasonal crannies from the predatory flames. The Earth lacked a fire broker, a creature capable of reconciling fuel's supply and flame's demand. The epoch's colossal stockpiling of carbon tracks a fire deficit so vast it had implications for the global climate, as much as its fire-catalyzed release does today.

That fact is, not all the elements of First Fire's informing triangle had been brought under biological control. Until that happened, fire could synthesize fuel and oxygen only spottily, if brilliantly. Those historic gaps ended with the arrival of fire-wielding hominids, who first made spark as steadfast as air and then readied fuels for the flame, and in fact did not limit their quest to fuel-foraging but planted and slashed what they wanted to burn and even ripped additional combustibles out of ancient rocks. Although carbon continues to recycle—charcoal from free-burning fire is one of the few mechanisms for shifting black carbon from the biosphere directly into the lithosphere—the dominant story today is the reverse: not storage but removal. Humans have exhumed fossil biomass and are burning it on such an immense scale that combustion and fire regimes now extend across geologic time. What failed to burn in the old Earth is burning in the modern. The limits on fire have increasingly become only those imposed by human will.

How Life Accommodated Fire

Life can exist without fire. The oceans prove that. But fire cannot exist without the living world. The chemistry of combustion has progressively embedded itself within a biology of burning. As life has evolved, so has flame: fire's history shows the same directions, drifts, and quirks as terrestrial life overall. Fire has prowled through the landscape of Earth history as a bear might search out berries, grubs, and fish; roaming or hibernating with the seasons, growing fat and thin with the yearly offerings.

The mere fact that fire exists has meant that life has had to accommodate it. Those organisms that could not adapt have suffered, those that could tolerate it have survived, and those that have discovered relative advantages in a burned site have found themselves oddly dependent on fire's regular return. In brief, fire has become a selective force and an ecological factor that guides evolution, organizes biotas, and bonds the physical world to the biological. It is as specific as the geotrophic orchids at the Cape of Good Hope that blossom within 48 hours after a burn and the Swedish beetles equipped with infrared-sensing organs that search out smoking stumps as nests. And it is as universal as the planetary cycling of carbon and the greenhouse gases sprung from combusted wood. Like storms and earthquakes, it disturbs sites; like fungi and termites, it recycles dead biomass; like sun and rain, birds and beetles, it is simply there.

What a History Needs to Know

Fire ecology is far from being a laboratory science. Control over the swarm of variables is weak, and field trials often fail to reproduce precisely the same results. The arguments for adaptation to fire largely belong in the realm of common sense and philosophy: an accommodation had to occur. The science of fire ecology is still struggling to document how this has happened. While these are serious qualms for fire sciences, they are less so for a brief history of fire. What matters are a few principles to guide our interpretation of what fire's history means.

First, fire does biologically what human ceremonies have unfailingly declared it to do: it promotes and it purges. It shakes and bakes. Around its flames revolves an ecological triangle, a circulation of biochemicals, species, and communities. It stirs molecules, organisms, landscapes. It kills plants, breaks down ecological structures, sets molecules adrift, shuffles species, opens up niches, and for a time rewires the flow of energy and nutrients. Fire upsets, shreds, reorganizes, revives, and quickens.

Second, plants and animals "adapt" not to fire as a principle but to *particular patterns* of fires. Even a single place may experience a variety of burns—a surface fire in the spring that lightly brushes off a veneer of needles, a wrenching crown fire in the late summer that guts a conifer forest. A biota can respond to a single event; it adapts to recurring ones. Over time, just such a pattern typically emerges, a mix and rhythm of burning that warrants the term "fire regime." Organisms adapt to those regimes.

Third, adaptation rarely takes the form of a single trait. Because fire occurs within contexts—a chemical environment that governs combustion, a physical environment that directs its behavior, a biological environment that shapes its ecology—adaptation is also relative to that complex. There are, to be sure, organisms that display traits that are apparently, and spectacularly, specific to fire. But more typical is a suite of traits that adapt the plant or animal to the range of conditions within which fire occurs. In brief, fire is one of the Earth's great interactive biotechnologies. Rarely has it occurred without drought or browsers for example. A trait that serves one purpose may serve equally well for others.

Fourth, fire is as ecologically powerful removed as applied. If fire is a biological presence as important as sun and rain, then halting it has the same force as blocking off sunlight or shutting down rainfall. A biota that knows a rainfall regime of 30 inches a year will suffer if it gets only 5, and that in a handful of downpours. So it is with fire regimes. Adaptations good for one set of circumstances become worthless, even harmful, if those circumstances change. What this means for fire history is that there is no ecologically neutral position possible. Not having fire is no more natural or benign than having it. For humanity, whose biological identity derives from "keeping the flame," there is no way to avoid fire. Deciding to apply fire, remove it, or change its rhythms, all have biological consequences.

More on Adaptations

How do organisms relate to fire? To simplify matters, these accommodations take two general forms: those that protect against fires and those that promote (or exploit) fire as a means to help the organism survive competitively.

Protective traits. The easiest features to identify are those that shield an organism from fire's passage. Thick bark shelters a larch's cambium layer from girdling by heat. Fat, succulent leaves guard the reproductive organs of aloes and proteas. Lignotubers store nutrients and even water from which mallee may resprout after fire has removed its crown. So, likewise, buried bulbs thrust upward after a liberating burn and an aspen's below-ground rhizomal trunk sends up new shoots. In fact, the same feature may serve several needs. The native grasses of American prairies store most of their biomass in underground roots, ready to blast upward when conditions warrant—traits equally well suited to survive drought, grazing, and fire. Not all grasslands know this full complement; far northern grasses may spend their dormant season under snow. But most will be grazed, undergo a period of curing and drying, and will sooner or later burn. For prairie fauna, the presence of underground burrows may lead them to roots and away from surface aridity and flame.

Organisms also possess many traits that fit them to particular fire regimes. The ponderosa pine has thick bark, can withstand fire-excavated cavities in its bole, and prunes away lower branches as it matures, lofting its needles not only above competitors for sunlight but away from surface flames, an ideal growth scenario for a country full of frequent underburns. By contrast, jack pine grows in gregarious throngs, whole swaths of even-aged trees mottling the landscape on a huge scale. The close packing of the canopy makes each patch vulnerable to crown fire under the right conditions. When the site burns, as it eventually will, massive seeding in receptive ash ensures that the succeeding forest will regrow its predecessor.

And so it goes. Some long-lived trees like coastal Douglas-fir and Australian mountain ash have fire intervals on a scale of 400 to 700 years. Some grasses like African sourveld and American tallgrass prairie thrive under a regimen of near-annual firing. Most plants tolerate a mix of burning, or if fires fall outside their adaptive range, yield to those that can. Or, more provocatively, they evolve features that encourage the kind of fire to which they are best adapted. They move from self-protection against fire to a fiery self-promotion.

Promotional traits. This is an elusive concept, an awkward argument. Geotrophic orchids, cheat grasses, bracken, an infinity of fireweeds—all seize sites purged of competitors by fire. They are opportunists, avid to seed and sprout before others, quick to claim any opening. But they cannot hold as well as they can grasp. After a few years they are crowded out of a recovering burn. So questions arise. Does their adaptation go further than taking what is presented to them? Have they so evolved that they begin actively shaping those sites and fire regimes to their own advantage? Do plants adopt growth habits and chemical properties that encourage fire—fires that give the plant some selective edge over its neighbors because they need a particular kind of fire to flourish preferentially?

The Earth tantalizes with possible examples. If lodgepole pine often requires heating in order to melt the wax that seals its serotinous cones, has it also grown in such a way that crown fires recur? Has the Australian grasstree, which flowers after being burned, evolved such that it makes that obligatory fire more likely? Is the peculiar evolution of flammability in chamise, a Southern California shrub, a biotic preparation to self-immolation? (As the plant ages, the chemistry of its leaves veers to a higher proportion of volatiles, the ratio of dead to live branches increases, woody debris collects at the base, the ventilation of the waxy scrub reaches an ideal. When it burns, it incinerates not only its surface self but everything around it; and then it resprouts, splendid in its selfish isolation.) And what about the regeneration of Big Tree sequoias, seemingly dependent on sites scoured by the slow deep burns that consume fallen boles? Is the chemistry of its combustion somehow tied to the physiology of its seeds, linked by cycles of fire?

The argument is hard to prove. But then it is equally hard to disprove, for the concept of "adaptation" can be viciously circular. There is no necessary reason why organisms could not have evolved traits to stimulate fire, however; and the exquisite choreography that seems to link high-intensity burning, in particular, with preferential reseeding suggests strongly that it has. These relationships are not the outcome of mechanical causes and effects—a fiery chisel sculpting a marble biota—but a long evolutionary give-and-take in which fire is a vital catalyst. What is clear is that organisms can live with fire, and that not a few seem to thrive amid it.

How to Think about Fire Regimes

Flame flickers through space and time. Fire regimes are as lumpy as the biomass that stokes them. The land is a quilt of burnable biomass, some patches vast, some tiny, much of that organic matter available for burning, some not. Those patches vary over time, some changing according to predictable scenarios, some not. Places have their histories, and these are as fitful as their geographies. What matters is that fires burn within these patches and within more or less regular rhythms. If they burn through several patches, they assume the characteristics of each one in turn as they spread, and if they burn at different times, each event exhibits distinctive traits. Yet some patterns do emerge.

The simplest means to reveal a fire's regime (its size, frequency, timing, and intensity) is to consider the distribution of water within a patch's biomass. Living or dead, fuels exchange moisture with their surroundings. If they are too wet, they won't burn. A fuel's moisture—hence that portion of its biomass that is available as fuel—changes with the daily wave of temperature and relative humidity, with the weekly passage of air masses through the region, with seasonal or monsoonal shifts in aridity, and with the long rhythms of drought or deluge on the order of decades or centuries. Severe droughts may crumble even rainforests into burnable fuels. Unusual rains may grow forbs and grasses in normally fireproof deserts, now flush with fine fuels and ready for flame. Long-term fire records around the Pacific Ocean trace nicely the pulses of the El Niño–Southern Oscillation (ENSO). And all these rhythms, of course, ride on the deep swells of climate change, as the Earth slides into and out of glacial ages. Whole biomes may arise and disappear or migrate across continents in the process. Fire will rise, fall, and travel with them.

These wet-dry rhythms set the ecological cadence for fire regimes. A place must be sufficiently wet to grow fuels and sufficiently dry to allow them to burn. Each day thus shows a preferred burning period, each year has its fire season. South-facing slopes make fuels available differently than north-facing ones do; wetlands burn on different cycles than do wind-swept plateaus. Patterns of wetting and drying, not the totals, are what matter—peaks and pulses, not averages; a seasonality of precipitation, not temperature. If the rhythms of wetting and drying are regular, so are fires. If they appear erratically, so will fire. Still, places assume more or less predictable patterns of burning. The fire-ecology equivalent to climate is the fire regime.

The concept has many flaws. It stamps a statistical mean onto what in nature is highly variable, and it tends to ignore the exceptional and the unpredictable, which (as with weather) are the events that yield the greatest effects. But fire burns living biomass as its fuel, and thus shares the diversity, exuberance, and randomness of life. Free-burning fire is not encased in the boiler of a steam engine, driving ecological gears with mechanical regularity. Natural fire does not rekindle with the metronomic precision of an automotive spark plug. For all its tangled skeins, however, First Fire is in some ways simpler to unravel than Second Fire. Once fire bonded with humanity, it had also to respond to ideas, institutions, beliefs, trade, and taste as much as to winds and ravines. Anthropogenic fire has had to understand itself in ways natural fire never has.

First Fire Today

Humans have so thoroughly restructured fire on Earth that it is difficult to find truly natural fire regimes. Everywhere we have remade fuel and recast spark, and thus reordered fire. Even landscapes now empty of people bear the marks of our former tenancy, sometimes having coexisted if not coevolved with anthropogenic fire from their very origins. Experiments to restore a pure, lightning-driven regime have generally proved frustrating or have failed outright because the legacy of human history cannot be wiped away. Even "pristine" landscapes exist through an act of human will. The concept of natural is itself a human invention. It is not "natural" for humans to vacate a landscape. It is not "natural" for humans not to burn. Since at least the Holocene, it has not been "natural" for lightning-caused fire to burn, as the saying goes, wild and free.

Fire Islands

Yet some uninhabited islands do exist. Some sites relatively uninfluenced by direct human meddling have appeared over the past decades, and some landscapes (like portions of the boreal forest), while subject to human-kindled fire, remain under the larger influence of fuels and climate and retain a substantial fraction of their primordial fire identity. They bear witness to the power and inevitability of natural fire. And they suggest, by both contrast and competition, how anthropogenic fire has worked its own alchemy.

There are true islands that exhibit lightning-driven regimes—isles in northern Swedish lakes, sandy patches of Scots pine amid the cold swamp that covers much of western Siberia, forested mesas in Utah and Arizona. The larger Swedish isles show more fire than their adjacent smaller ones, probably by virtue of offering a larger target and sporting lightning-favored trees rather than brush. Still, the frequency of fire is small, measured in centuries. The Siberian pine patches suggest a fire interval of 60 to 70 years, which is still more frequent than the spruce-fir regime in surrounding areas. Remote basins—the biota-filled cavities of old cirque glaciers—in the Sierra Nevada and Rocky Mountains experience fire in splendid granitic isolation. A 1937 expedition to Shiva Temple, a 300-acre mesa in the heart of the Grand Canyon, noted "trees scarred by lightning and evidence of forest fires which had not done much damage." Here frequent surface burns had fashioned a classic pine savanna.*

Those scenes simply confirm the obvious: that fire exists independently of humanity, that biomes and lightning fire reach some sort of accommodation. It is not a simple task to extrapolate further. Such sites suffer the liabilities of island biology, a serious concern because fires, like migratory species, move. Most places burn not because lightning has ignited a tree on that precise site but because fires burn into them or across them. Islands, particularly small islands, prevent this. They remove from the scene the most dynamic of fire's properties, its capacity to roam. This fact accounts also for much of the fire problems associated with "biotic islands" such as nature reserves. Fires aren't allowed in, and are controlled before they can move out. If the reserve lies on a mountain, the effect is worsened because the site is cut off from fires that would, historically, have climbed into it from the lower, drier, sooner-available fuels in valleys below.

First Fire's Reserves

Bigger reserves generally overcome this difficulty, though not completely. Scale alone is not decisive: no place is large enough to capture all possible fires, and none probably large enough to absorb the very biggest burns. But fire size is only part of what shapes a fire regime. Inviolate reserves, in particular, often deny a place's fire history. They pretend that anthropogenic fire was unimportant or that lightning can find a quick equilibrium in a system that humans have long sculpted either by starting fires or by putting them out. In removing people, nature reserves have remade fire's context in often unpredictable ways. Dumping fire—lightning fire least of all—into such a landscape may have little relation to First Fire ecology. It may take centuries for an "equilibrium" to emerge, if that term has any meaning over such long spans of time.

*David A. Wardle et al., "The Influence of Island Area on Ecosystem Properties," *Science* 277 (29 August 1997): 1296–1299; S. N. Sannikov and J. G. Goldammer, "Fire Ecology of Pine Forests of Northern Eurasia," in Johann George Goldammer and Valentin V. Furyaev, eds., *Fire in Ecosystems of Boreal Eurasia* (Boston: Kluwer Academic Publishers, 1996), p. 152; Harold E. Anthony, "The Facts About Shiva," *Natural History* 40(5) (December 1937): 718. See also David Parsons and Jan van Wagtendonk, "Fire Research in the Sierra Nevada," in William L. Halvorson and Gary E. Davis, eds., *Science and Ecosystem Management in the National Parks* (Tucson: University of Arizona Press, 1996), pp. 35–46.

Still, a small but growing body of evidence suggests how natural fire might work. In California's Sierra Nevada, two national parks have, since the late 1960s, evolved programs to promote a lightning-dominated fire regime. The experience confirms what most researchers had intuited or deduced about natural burning: that a few major fire years account for most of the burned area and do most of the biological work, that fires burn in patches which in complex ways check one another, and that fire in the larger perspective is conservative, acting to maintain what exists. It is not clear whether such a regime can expand meaningfully over more of the park (probably not); whether it could exist without prior preparations and monitoring (not likely); and whether, if successful, it is sufficient to maintain such special species as the Big Tree groves (such an experiment is not likely to be tried). These lands have coexisted well beyond the life of even *Sequoia gigantea* with people and their fires. The odds are at least even that abolishing these practices might abolish the trees as well. Larger preserves such as America's Yellowstone National Park and the Kruger National Park in South Africa encompass immense areas, sufficient to overcome some island effects, but introduce problems of human history and philosophy that muddy the meaning of their experiments.

A useful compromise is the Gila Wilderness in the mountains of southwestern New Mexico. Lightning fire is abundant, annual, and ancient. Natural fire exists, in some form, virtually every year. In 1972 the reserve adopted natural fire as a goal, and began to absorb more and more ignitions. The key to the regime's dynamics is the temporary uncoupling of wet and dry conditions: the big fires occur early in the spring, the major lightning (and rain) in the late summer; the big fire years occur during drought years after one or more abnormally wet years. The record of fire, chronicled in the tree rings, clearly tracks the regional climate.

Even this may be an artifact of human history. While the role of anthropogenic fire remains ambiguous, the pre-European peoples practiced an aboriginal economy for which fire was vital. They controlled ignition, which in this kind of fire regime is enough, for it is possible for people to seize mastery with little more than firesticks. By keeping fire constantly on the landscape, however, they have left climate as the principal variable. Climate looms exceptionally large in the record, and its changes are notably important, because the spark is always there. End that unquenchable flame and the historical record might look different. In fact, the 20th century has removed those torches (along with much of the grassy understory that carried flame), and as a result, fire history reflects the political economy of ranching and forestry as much as it does the spasms of ENSO. (Oddly, those who argue that the record of fire-scarred trees tracks only natural fire grant humanity's capacity to alter those regimes by suppressing flame, which is difficult, but not by starting it, which is easy.)

In sum, what we see today is not wholly natural. How could it be? What we have, however, suggests powerfully how First Fire works and what its regimes might look like.

TOUCHED BY FIRE

With Homo erectus, *the biosphere began to influence ignition as it did oxygen and fuels. Fire's triangle was becoming organic, though at a cost. All photosynthesizers contributed to atmospheric oxygen; all terrestrial plants were, at least potentially, available as fuel; but only one species controlled ignition. Nature's economy had found a fire broker, then granted it a monopoly. That species' fire-leveraged power over ecosystems was enormous.*

All humans manipulate fire, and only humans do so. We are truly a species touched by fire. Fire opened the night by providing light and heat. It protected caves and shelters. It rendered foods more edible, leached away toxins from cassava and tannic acid from acorns, and killed bacteria that caused salmonella, parasites that led to trichinosis, and waterborne microbes. It interacted with every conceivable technology from flint mining to ochre painting. Fire was a god, or at least theophany; fire was myth; fire was science; fire was power. We could call it forth as we could not call forth floods or hurricanes or earthquakes or droughts.

The control of fire reformed hominids as well. It changed diets. It made food accessible that otherwise was too toxic or tough to consume. It released the skull from having to brace the enormous muscles required to chew uncooked foods, thus perhaps allowing the skull to swell and the brain along with it. Certainly fire's possession altered social relationships. Groups defined themselves by their shared fire. Domestication itself most likely began with the tending of flame. Like a being, it had to be conceived, fed, protected, put to bed, awakened, trained, controlled, exercised, bred—in effect, socialized into human life. It required constant attention. It needed a protective shelter (a domus, *from which comes "domestication"). Someone had to gather the endless fuel, someone had to fuss over the flames and nurture the coals, and someone had to oversee its proper use. For it to expire was a calamity.*

But was fire really that critical? Try this, as a thought experiment. Remove every vestige of tamed fire and examine what remains. Remove the hearth fire, the cooking fire, the protective ring of evening flames, the fires that softened flint and hardened wood, the fires around which humans gathered to talk and share stories and learn about the meaning of the stars and the grassy veld. Remove that central flame and the center no longer holds. Humanity's other tools, those shaped without recourse to controlled fire, grant humans no greater power than the talons and fangs and bulking muscles and sense of smell that our competitors possess. As fire myths so universally declare, without fire humanity sinks to a status of near helplessness, a plump chimp with a scraping stone and digging stick, hiding from the night's terrors, crowding into minor biotic niches.

But the Faustian pact with fire was reciprocal. If fire freed humanity, it is also true that humanity unshackled flame. Every place humans visited, they touched with fire. They brought fire to landscapes that had not known it. They restructured fire regimes that had experienced one kind of fire and gave them another. A wet-dry cycle worked on biotas like a frost-thaw cycle on rock, cracking open the landscape and allowing humans to drive their fire wedge into the fissures. Fire and humanity pushed and pulled each other around the globe. They advanced together—spreading like flaming fronts, spotting into favorable sites, probing into marshes, flaring amid thickets, smoldering amid peat, crackling through scrub, all as the fuels of environmental opportunity and the climate of culture allowed. Charcoal is among the most reliable of records of this hominid diaspora. The residue of the hearth fire, the charred bones of meals and cremations, a site-shattering layer of black carbon that marks a dramatic shift in the ecological order are all the signatures of human passage. In real as well as symbolic terms, humanity and fire had come to resemble one another such that the tread of one tracked the tread of the other.

In The Republic *the philosopher Plato likened the human condition to life in a cave, illuminated only by flames. But the allegory is deeper than Platonic idealism. In Swartkrans, a South African cave, the oldest deposits hold caches of bones, the prey of local carnivores. Those gnawed bones contain the abundant remains of ancient hominids. Above that record rests, like a crack of doom, a stratum of charcoal; and atop that burned break, the proportion of bones abruptly reverses. Above the charcoal, the prey have become predators. Hominids have claimed the cave, remade it with fire, and now rule. That, in a nutshell, is what has occurred throughout the Earth. What has happened with early prey relationships happened also with fire. As humans successfully challenged lightning for control over ignition, the whole world has become a hominid cave, illuminated, protected, nurtured, warmed, and controlled by the flame over which humanity exercises its unique power and through which it has sought an ethic to reconcile that power with responsibility. Ours became the dominant fire regimes of the planet.*

Yet the Earth did not get quite what it supposed. The biosphere needed a reliable spark whose timing obeyed biotic rhythms, subject to ecological processes, shaped by natural selection. Ideally ignition could be coded by instinct. A creature would set fires much as elms shed leaves or salmon turned upstream to spawn. What nature got instead was a sentient being whose neural net was short-circuited by synapses of society and culture. The Earth's keeper of the flame kept it for his own purposes.

Chapter Two

Frontiers of Fire (Part 1)

FIRE COLONIZING BY HOMINIDS

Humans brought to the Earth what the First Fire landscapes of the Devonian and fuel-surplused landscapes of the Paleozoic and Mesozoic had lacked: a creature who could carry fire around its surface, who could match fuel with flame. What began as a chemical event evolved, in humanity's restless hands, into a device for remaking whole landscapes. No human society has lacked fire, and none has failed to alter the fire regimes of the lands it encountered. Equipped with fire, people colonized the Earth. Carried by humans, so did fire.

Every part of the fire triangle proved pliable. Humans could start fires at eccentric moments and with odd timings, halt unwanted fires, begin burns under circumstances that made them large or small, hot or cool, that forced them to back down slopes or let them billow with the wind. Not every spark cast ended in a flame or every fire in a flaming front, but ignition became as constant as human will desired. Less easily, we have learned to tweak biomass into available fuel. We can add species—sometimes more fire-prone, sometimes less. We can build up and tear down fuel loads, restructure them, make them easier to fan or tougher to dry. We can cut, prune, log. We can stock domestic grazers and browsers, and kill off their wild competitors. We can plow, harrow, plant, weed, harvest, and fallow. We can even modify local weather. We can reset wet-dry cycles by irrigating and draining, and by ring-barking—killing and parching—dense forests. We can tinker with microclimates by altering the sunlight that strikes the surface, the ease with which winds can blow through woods, the ability of the land to reflect light, its capacity to hold or shed moisture. We can erect wholly new fuel arrays (fire habitats, if you will) in the form of houses and towns. Anything that modifies the vegetative cover influences how fire will burn. All this is within our reach, never more so than when we grasp a torch.

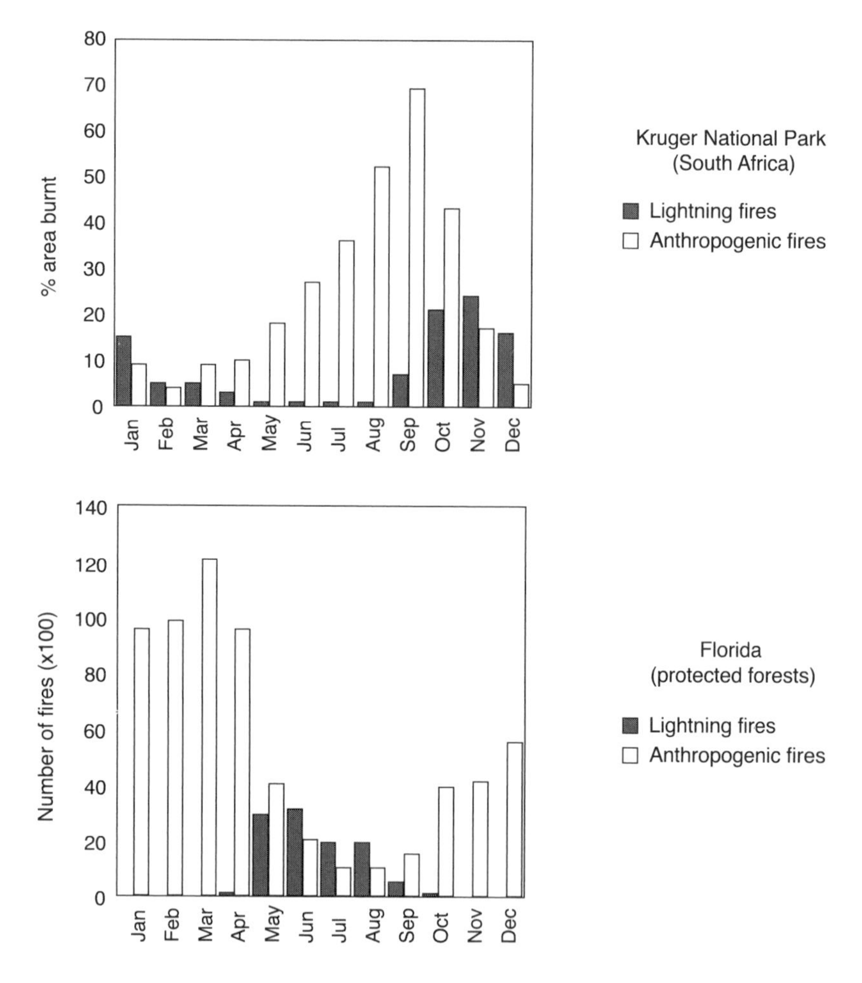

FIGURE 2. Two seasons, two competing fires. There is little basis for the popular belief that pre-agricultural peoples use fire according to the same calendar as lightning. Rather, the almost universal pattern is to burn prior to the lightning-fire season. As more fuels dry out, they are burned. By the time lightning arrives, most of the burning, done either to promote some feature or to protect a site, is complete. This pattern continues even into modern times.

Consider two contemporary examples. Kruger National Park (South Africa; top) has a pronounced lightning-fire season, but most of the burning occurs outside it, expanding with the buildup of available fuels. While there is a period of overlap, most of the burning occurs through human hands and during the prolonged drying that precedes the rains. Florida's protected forests (bottom) experience the greatest barrage of lightning of any woodlands in North America and kindle more frequently from that source than anywhere east of the Rocky Mountains. But even here, and even with routine human manipulation excluded, the majority of fires start from people and burn outside the lightning season. (Sources: Trollope et al. 1995 and Komarek 1964, both redrawn by the University of Wisconsin Cartographic Lab)

What Made Early Fires Effective

The principles behind such powers are simple enough. First, people with limited technology are more effective in landscapes that already have fire than in those that lack it. The fuels are on the ground and the biotas well seasoned by flame. Equally as good, but less common, are places that have heaps of fuel and a suitable rhythm of wetting and drying yet lack ignition. In either site all one needs is a controlled spark to take over the fire regime. Early hominids outfitted with throwing sticks, wood drills, flint points, and scrapers could thus unhinge whole ecosystems, provided they could wrestle their firesticks onto a suitable fulcrum. Modern city-dwellers shun fire-prone places as dangerous, yet their ancestors sought out fire-prone landscapes precisely because those places *could* burn and thereby granted humans power over them.

The stickier landscapes are those for which fire is rare, for which the controlled spark fails because nothing burnable exists for it to strike. Such places require control over fuels, which typically hinges on having tools to puncture and pry apart the flora and fauna sufficiently for sun, wind, and fire to enter. Goats and hogs may be as effective in this task as axes and saws. Either way people redesign the landscape itself to accept fire. Not every society could do this; probably none before the advent of agriculture.

But how much change could a shift in timing make? Is the shuffled scene a matter of degree or of kind? Did early humans only modify what existed—pruning a biome that was already there—or did they create something new? Our fire practices, after all, derive from nature. We did not invent fire: we took (borrowed, stole, fought for, connived) fire from lightning. So it is with fire hunting and fire foraging and slash-and-burn farming, all of which originated out of natural models. There is a sense, then, that early hominids only enlarged the existing dominion of fire, that they merely helped train what nature had already bred.

Yet there is little evidence that people have sought only to complement natural fire. Close study argues rather that early humans (and aborigines of similar economies) actively competed with lightning fire. Only so much biomass existed. What people did not or could not burn, lightning would, and if nature burned away those fuels, then firesticks could not work much magic. The simplest solution was for people to burn that fuel first, and around the world similar patterns of early burning stand out. Anthropogenic fires begin as soon as swaths of fuel cure. They continue through the dry season, multiplying and spreading as more and more fuels ripen. The fired parcels mesh one with another, typically growing larger as the season lengthens, maturing into a mosaic of burned patches and corridors. If lightning arrives, it can burn only those combustibles that remain. The critical landscapes—those most valued, those most needing shelter—have already passed through the protective flames or are shielded by them. In this way, fire's regime changes as people nudge and heave the timing and siting of flame, even in those landscapes that regularly burned before humans arrived. But the *regime* is what the plants and animals adapt to.

We are fire creatures from an ice age. Our ancestors matured rapidly during the alternating climatic currents that sloshed through the Pleistocene. For more than two million years, the Earth swung between glacial and interglacial, pluvial and interpluvial, between cold and warm, wet and dry. Some places sank under ice and water, others dried and became windblown. Forests and grasslands ebbed back and forth over landscapes like vast tides. These are cycles that, at a faster tempo, favor fire. On the scale of the Pleistocene's long swells they favored a fire creature.

Probably *Homo erectus* emerged in Africa between 1.5 and 2.0 million years ago. By 1,000,000 B.P., certainly by 500,000, the species had reached China and Java and likely most other places in the Old World that were not submerged under glaciers or seas. Possibly other hominids appeared as well, among them the Neandertals. Then sometime between the last two glacials, between 100,000 and 150,000 B.P., anatomically modern humans, *Homo sapiens,* emerged and again swept out like a flaming front, this time everywhere.

To a remarkable degree, that fire analogy is literally true. Charcoal is the spoor of early hominids. The record of their wanderings is preserved and dated by fire. Hearths mark the caves they occupied, the hide huts and wattle windbreaks they erected, the sites where they hunted, butchered, and cooked mammoths, bison, and boars. Charring shielded the wood and bone against decay (and allows for carbon-14 dating). Burned relics identify the site as the unique, if temporary, residence for the one species that could so ply fire. Early hominids left behind fossil fire as they did flint flakes and drilled bone.

Yet fire was different from stone scrapers and wooden clubs. Almost virus-like it could pervade whole landscapes, reset the rhythms of the seasons, reorder flora and fauna. And it could interact with other tools. Much as an atlatl adds thrust to a javelin, so fire could add ecological heft to spears, axes, and snares. Burning and hunting, for example, had impacts greater than either practice alone. Controlling the populations of herbivores by hunting, flushing up or burning off their limiting ranges, or wiping out their predators can profoundly influence the vegetative cover, which is to say, the fuels that feed fire. The upshot may be more fire (or less) but the scene is unlikely to stay the same.

The astonishing scattering of humans over the globe coincides with a wave of megafaunal extinctions that deeply implicates fire. Most likely Second Fire helped prod those extinctions; certainly it found a different world after those creatures had departed. Of course the Earth had suffered biotic binges and busts, including extinctions, on a huge scale before humans arrived. The immense climatic fluxes of the Pleistocene had, quite without human aid, wiped out thousands of species. In one sense, humanity was itself simply a part of a cycle of megafaunal recolonization. Yet it is also clear, from historical evidence—melancholy examples from the dodo to the moa—that humans trekking into new lands can quicken natural trends and can push species to the brink. Much as they herded mammoths into bogs at Torralba or bison over canyon rims at Head Smashed In Buffalo Jump, so we have driven scores of species over an evolutionary cliff. Were the Pleistocene extinctions the last of an ancient natural lineage, or the first of a new order of hominid rule?

There is no definitive answer, for too much was changing. Climates roamed like storms over whole continents. Lands were flooded by rising seas, and landscapes were exhumed from receding ice and evaporating lakes. A biotic scramble ensued to claim these remade worlds. Whatever humans did they did within an era of dramatic change, one that favored an omnivorous, wandering, curious, wily, and adaptable creature, especially one that could wield fire because what mastodons and woolly rhinos had once eaten might now be available to feed flame. In the end, the Pleistocene's megafaunal menagerie was replaced by humans and, in select lands, by their tame livestock. Certainly people influenced that exchange, undoubtedly with the aid of fire. For extinction, it is not enough to slay animals. Their return must be prevented, and that is where burning went beyond a device to help hunt. Together, spear and torch remade a host of habitats on scales both tiny and vast.

Those big animals had mattered. They bashed, trampled, selectively gobbled up or spared, and rearranged a scene's fuels. Even today, herbivores (and the carnivores that prey on them) profoundly shape African fire regimes. What the antelopes and wildebeests eat, fire can't burn. If elephants don't bash over trees and rip off branches, shade crowds into the land and fire must struggle to survive. How fire and hunting interact depends of course on local conditions; foremost, on the innate fire-proneness of the land. Wipe out megafauna in a place with a pronounced wet-dry cycle, and you can keep that landscape open and grassy, or mottled with brush, through regular burning. If anything, fires become more powerful because there is more to burn; the bulk eaters no longer compete with flame for grass and browse. Granted these simple dynamics, it is astonishing that closed, fire-free forests exist anywhere.

The explanation is that if you clear out those same creatures in a scene in which a seasonality of wet and dry has broken down, the fires are likely to fade away. Torch and spear alone cannot fight back a maturing shade forest. Against the growing damp and the dark, the flames falter, and like a candle under a bell jar, slowly expire. Without something to crack open unflaggingly wet woods, flames simply flicker out along their moist margins. In such places, people with tame herds could substitute their cattle, sheep, goats, and hogs for the lost megafauna. These could chew, snip, and trample through the scrub. Probably, too, those servant species worked with servant fire, which their human herders freely

FIGURE 3. First-contact fire. People and fire arrive together. If fire already flourishes, the effect is to shift the fire regime. If the conditions for fire exist but fire does not because there is no consistent ignition source, however, then the arrival of humans can create a dramatic shift in the biota and leave a signature layer of charcoal in the soils and the sediments of lake beds.

Consider two examples from sediment cores. The first (top) derives from Lynch's Crater in Queensland, Australia. This simplified summary of the core shows a relatively abrupt spike of charcoal at 39,000 years BP, which coincides with a biotic adjustment (as recorded by preserved pollen) in which tougher, more disturbance-adapted species replace the long-reigning "rainforest" species. This date corresponds locally with the first remains of humans. The second (bottom) comes from a core taken from Lake La Yeguada in Panama. It tracks a sudden shift in soil, plants, and charcoal, and agrees in chronology with evidence for first human contact. The most likely explanation is that changing geographic conditions made fire possible and people kindled the spark. (Sources: Clark 1981 and Colinvaux 1997, both redrawn by the University of Wisconsin Cartographic Lab)

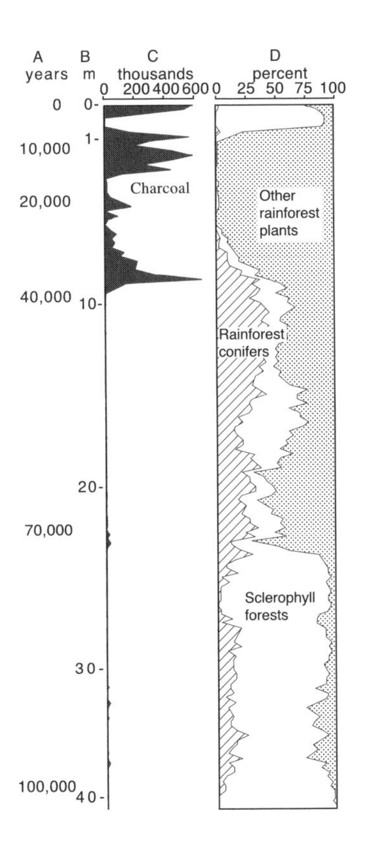
A
years
B
m
C
thousands
0 200 400 600
D
percent
0 25 50 75 100
0
10,000
20,000
40,000
70,000
100,000
0-
1-
10-
20-
30-
40-
Charcoal
Other
rainforest
plants
Rainforest
conifers
Sclerophyll
forests

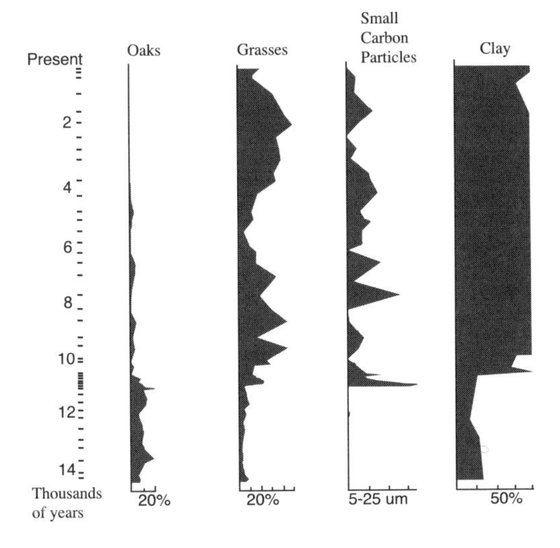
Present
2
4
6
8
10
12
14
Thousands
of years
Oaks
20%
Grasses
20%
Small
Carbon
Particles
5-25 um
Clay
50%

granted. Without such aid, however, people, like their fires, had to retreat to more open fringes. They had to flee to riverbanks and patches of heath, baked-soil savanna, and sandy barrens.

Whether or not humans caused or assisted those Pleistocene extinctions, they thrived amid the turmoil of those times, and they colonized landscapes jolted by those epochal losses. They went nowhere without flame. It was their great enabler. It migrated with them as tool, servant, camp follower; as agent, accident, and index. Where one went, so did the other, which meant that both traveled very widely indeed. The geography of earthly fire looked vastly different when the Pleistocene ended than when it began.

First Contact: When Fire Arrives

How anthropogenic fire—Second Fire—struck a land depended on local conditions, on timing, and on what else humans carried in their toolkits. Did it meet a land whose seasonality of wet and dry was sharpening or blurring? Was it a land that was shedding forests, as a bear would its winter coat, or piling up peat like seal blubber? Did fire's blows strike stone or wood or grass? Did it hit directly or through the chisel of herding and hunting? Did flame find good fodder, or did it have to wait for fire-hardy weeds to take over? There emerged, in short, a gradient of burning as there was a gradient of megafaunal extinctions. First-contact fire stories are many.

The primary consideration is whether fire is already on hand or not. Then it matters how anthropogenic fire arrives, whether it is part of a long chain of human burning or whether it comes, ecologically speaking, as a bolt from the blue. In most of the Old World—Africa, Eurasia—anthropogenic fire advanced bit by bit. The torch passed from one hominid to another; landscapes already adapted to one regimen of human-wrought fire accepted another. Fire's power steadily improved in step with the rest of the hominid toolkit: each addition increased the other's leverage, so that fire could do more and people could apply flame to more purposes and with greater effect. It advanced and retreated and then advanced again. Mostly though, Second Fire toughened its presence, probed further into fire-friendly fringes, strengthened its ecological authority.

But some places—the Americas, Australia, islands large and small—

escaped those Old World preliminaries and plunged directly into contact with modern humans and their blazes. There are vast biotic differences of course between New Guinea and Greenland, Easter Island and Mauritius, New Zealand and Iceland, and chasms no less profound divide their encounters with fire. The critical geographic question is how fire-prone they were. The critical historical question is if their first-colonizing peoples had agriculture or not. Whether the far-flung sparks of contact kindled depended on exactly where they landed and what assistance people could give them by way of stoking the feeble flames. Some of them flared, many expired. Some places snuffed out embers as fast as their colonizers could blow life into them and, long unshaken by fire, could be made fire-prone only by first slashing their forests into tinder or setting herds to crunch through the scrub. Others already reveled in routine disturbances, readily ceded their megafauna, and knew free-burning fire as a birthright. Like many creation myths, their first-contact stories begin with world-shattering fires.

Colonizing Fire-Prone Places

On fire-prone lands, all colonizers needed was a well-sited fulcrum; often a seasonality suitable to fire was enough. It was easy for people to preempt lightning—they had only to burn off the valued landscapes before the seasonal storms arrived. If they could also eliminate megafaunal competitors for biomass, then there was that much more fuel available. With little more than a firestick, aboriginal societies could move whole landscapes.

Surely the most spectacular instance is Australia. Here Aborigines equipped with torches, spears, and throwing sticks moved a continent. Like other peoples, they had every incentive to do so; unlike others, they had conditions favorable to making it happen. The places richest in resources were prone to burning. The southeast and southwest corners have mediterranean climates, with short, wet winters and long, dry summers, both subject to drought and plumped with fuels, yet lean of lightning. The northern tier washes in the tides of the Asian monsoon, with well-defined wet and dry seasons, and crackles with lightning at the onset of the wet season rains. Here people could snatch fire away from nature easily. Most of the rest of the continent relied on long waves of drought and deluge that promoted burning if there was a reliable spark. Even the megafauna—the other competitors for biomass—largely crumpled on contact. What resulted was a place ripe to burn.

Fire flared widely, and stayed. There are sharp spikes in the charcoal chronicle left in soils and lakebeds that coincide roughly with human contact and the disappearance of many megafauna. Did Aborigines kindle those fires? Of course they did. All peoples do. The question is whether the fires could spread, and if so, how extensively they could sway the overall biota. It seems likely that much of Australia was primed for fire, awaiting only a spark. So was Aboriginal burning a cause or a catalyst, or do those distinctions really matter? The mix of plants and animals shifted massively and suddenly in Pleistocene Australia at the same time that fire broke out as a chronic, continental presence. It was the precise moment that people arrived in force. Whether Aboriginal fire caused those conditions or merely seized upon them, the colonization of Australia is a prime testimony for the power of first-contact fire.

Colonizing Fire-Intolerant Places

Yet there were many places that did not have flame or the circumstances to promote it. Fire's arrival under such terms was tepid and tentative. For humans to thrive, fire had to thrive. Yet for fire to flourish, colonizers had to create a habitat for it, and not all peoples came equipped to do so. Trailing their firesticks, the Andaman Islanders, for example, proved unable to fracture the dense woods sufficiently to support broadcast burning. So they, like their fires, remained in canoe and house, huddled along the coast, and poked into the interior along dark paths. While they attributed to fire the greatest of powers, it remained a lesser god in a world that did not know combustion, its survival as marginal as that of its tenders. In this the Andamanders speak for all the aboriginal peoples forced to fire's fringes.

What typically breached such fire-intolerant landscapes was agriculture. Farmers chopped, stacked, planted, watered, and left to fallow—practices that converted biomass into fuel. They could set wet woods to dry, drain swamps, open canopies, sow and nurture new plants. They could also turn out livestock to chew and root up and reseed. Like Noah's, the arks of many settlers bulged with beasts, two or more of each, ready to multiply. After all this fussing and all these beasts had pawed and chewed over the landscape, colonizers could burn, and did. But perhaps more to the point, they invented new reasons to burn. They did not seize and redirect an existing fire regime so much as cultivate a new one.

Where it did not exist, they could grow fire as they did wheat and melons. When agriculture furnished the fires of first contact, the outcome could be dramatic.

But agriculture complicates the narrative, because only the finest of lines separates colonizing from cultivating. (The latter is, in a sense, the domesticated version of the former.) Agriculture can thus fire-colonize almost any landscape. It can remake aboriginal landscapes as well as uninhabited ones. It has its own sagas of colonization. Yet there is a value to considering such stories, for they underscore what makes fire-colonizing work. They sharpen our understanding of why some peoples could shatter lands while others could not, why fire could smelt some landscapes and not others. One of the most interesting of these stories tells of Europe as it groped across the Atlantic islands.

The core practice, now termed *landnam*—literally, "land taking" in old Norse (pronounced *landnahm*)—shattered the fire-free shade forests that covered central Europe. Pioneers packing seeds and goats as well as axes and torches soon spread from the mainland to Europe's peripheral islands, and to the igneous chunks that studded the Mediterranean, the skerries rimming the northern seas, the microcontinental isles that hopped across the Atlantic. All became littered with the charcoal of landnam, and some acquired eternal flames. The uninhabited islands within the pale of the Greater Mediterranean—the Canaries, the Azores, Madeira—were easy targets. Their climate was similar to the inhabited regions, their flora derived from common ancestral sources. Where they differed was that they lacked humans, agriculture (especially livestock), and fire. Colonists brought them all. Often voyagers would drop off some sheep or goats on a newly found isle to sap the foundation of its tough scrub, easing the transition to fire. When they returned, they soon forged the isles into Mediterranean miniatures.

Across the North Atlantic, Norse voyagers bumped into a very different landscape, one on the habitable margin. The scenes, however, did not differ dramatically from those of coastal Scandinavia, where landnam had broken woodlands into heath. When Viking colonists landed on the Faeroes, the Shetlands, Iceland, and Greenland, they applied the full force of continental landnam. The forests regrew slowly, if at all, so first contact failed to renew itself into long-fallow farming. Settlers instead turned to flocks and fish. Still, those hard-worked soils carry the charred archives of fire's first contact.

Mythologizing First-Contact Fires

It is particularly true for agriculturalists: the saga of first contact takes the form of a great fire. The Malagasy called it *afotroa.* Maori myths record how the first arrivals lit fires everywhere, burned off forests, and wiped out moas. Madeirans preserved the legend of a Seven-Year Fire that drove the first settlers into the sea for protection and then, smoldering, left the isle as malleable as a lump of white iron drawn from a furnace. The cosmology of the Stoics was built around a recurring world conflagration. The Aztecs performed a New Fire ceremony, symbolically rekindling the world, every 52 years. Modern mythmaking has continued the trope. Star Trek's *Wrath of Khan* features a "genesis device" capable of remaking whole planets. The "genesis effect" begins with a fiery blast and spreads its "new matrix" over cold-dead rock with a flaming front. More slowly and more bumptiously, that is precisely what humans did with the Earth.

Even places without the raw stuff for cosmic conflagration could find a surrogate. Iceland, for example, lacked the thick-wooded fuels to stoke a Ragnarok-sized world burn, however hypothetical. Yet the memory of its founding practices endured, not only in char-laden earth but in the lines of the *Landnamobok* (*Book of Settlements*), the written register of landnam. When, after sixty years of chaotic scramble, Icelanders needed to reestablish land title in a methodical way, they reenacted the means by which they had first taken possession. For a day each landowner lit great bonfires and then walked, torch in hand, as far as he could advance from dawn to dusk. The lands he symbolically burned would be those he could hold in title.

So, with fire in hand, had humans always laid claim to new land.

Lost Contact: When Fire Departs

First-contact stories can also be told in reverse. When people leave, they take their fires with them, along with all their pruning, shuffling, foraging, and other ecological fidgeting. But because fire is as potent removed as applied, both tales have great significance for understanding how fire works.

Whether people come or go, the critical question is whether the land can have fire on its own or not. In abandoned lands that are prone to fire, lightning may reclaim the scene, such that only fire's regime will change. But landscapes that burn solely because humans work hard at it may revert to unburnable lumps. In this case, as with first-contact stories, it matters how people had used fire, whether they practiced an aboriginal or an agricultural economy, and whether they had livestock or not.

How to Create a Fire Vacuum

Lands that are both vegetated and empty of people are rare. Yet, if only temporarily, it can happen through war, disease, famine, or sheer wanderlust. Italy during the Second Punic War, western France during the Hundred Years' War, most of Europe while the Black Death raged, southern Africa during the Mfecane (or "Time of Troubles") in the 1800s, a chunk of Ukraine when the Chernobyl nuclear plant belched out radiation—all emptied landscapes. All broke the pattern of fire people had laid down on the land. All reset fire regimes, either by destroying the source of their fuels or by defaulting to lightning for the timing of ignition.

Yet European expansion by sword and sickness into the Americas, Australia, and (somewhat differently) Siberia went beyond these garden-variety episodes. The demographic collapse was both rapid and profound. Worse ecologically in the Americas and Australia, it resulted in a faunal deficit. The people were gone; they left few (or never had any) livestock; the native megafauna from the Pleistocene had vanished with particular thoroughness. Altogether the collapse removed from the vegetation both fire rivals and fire allies. It would take time—a century, perhaps two—to repopulate the mega-mammals. Meanwhile, the land went to seed, not only growing different species but reorganizing them into new patterns. What had once fed fire or fauna now became fallow, much of it perhaps inedible or unburnable. A garden had become a wilderness.

Something like this seems to have occurred throughout the Americas. Without their human tenders, landscape after landscape went feral, and their fires either ran wild or expired. The habitats that swiddening humans had carved for fire disappeared, absorbed by the same jungle that overgrew stone ruins, overrun by a woody scrub that only torch and ax had held in check. The prairies that hunters had routinely flushed with fire sank beneath the infill of brush and trees. Corridors that speeded fire like fuses were snuffed out. Marine sediments off the Pacific coast of Mesoamerica suggest that the flux of fire-driven charcoal has never equaled the rates that existed prior to the Conquest. (Incredibly, contact so shattered some of the Amazonian tribes, like the Guaja, that they actually lost the art of fire making.) That story can probably stand as synecdoche for all of the Americas.*

Still, North America was exceptional. Here the colonizing-caused vacuum persisted: the creation of public lands, primarily in the Far West, interrupted the process of Second Fire recolonizing. Here one pattern of anthropogenic fire would not substitute for another. No one would inhabit these lands, not as habitation is traditionally understood. Instead a strategy of fire exclusion became the announced political goal. Livestock—cattle after sheep—crushed the fine fuels that had sustained the old regime, then government officials sought to extinguish lightning fires as well. On public lands like the national forests, even logging could not supply suitable fuels as rapidly as old ones disappeared. Slash was too local, and while plenty flammable, too easily protected when officials wished; mounds of large-diameter wood were no substitute for sweeping horizons of fine needles and grass. Instead, the landscape was reclaimed by Third Fire, which sought to purge all flame from the scene in favor of its own internal combustions.

When First Fire Returns

As anthropogenic fire waned, the relative power of lightning fire waxed. From the Rocky Mountains to the Pacific Coast, lightning began reforging fire regimes on a near-continental scale and with such dominance (in some places accounting for 90 percent or more of all ignitions) that officials came, in time, to doubt the former strength of the dwindling native peoples or even to see the biotic power of their once-ceaseless burning. Officials and intellectuals no longer recognized in this overgrown landscape the legacy of Second Fire. They no longer witnessed routine controlled burning, and came to believe that the lightning-driven fire regimes they currently saw must have always prevailed, that the preserves they administered were relics of true wilderness, not artifacts of a historical accident. They knew that fires were missing, and assumed that their fevered suppression of lightning-kindled fires was the cause. They did not appreciate that the missing fires might be anthropogenic, or that fire's passing might prove ironic and baleful, much less that fire's slow death could inspire a violent rebirth.

*See William Balée, "Indigenous History and Amazonian Biodiversity," in Harold K. Steen and Richard P. Tucker, eds., *Changing Tropical Forests* (Durham, N.C.: Forest History Society, 1992), p. 193.

What stressed such assumptions to the breaking point, however, was the creation of nature reserves on scales that mock the old tradition of sacred groves. The fire history of these reserves shines perhaps most revealingly in the United States. Here a national creation myth found expression in political institutions to create, strengthen, and purify wilderness reserves. Once established, these sites knew little anthropogenic fire; rather, fire protection—ideally, fire exclusion, to save the sites "from fire and ax"—had been a founding goal. By the late 1960s, however, attitudes had changed. Fire suppression itself seemed intrusive, wrong in both its biology and its ideas. Rather, fire, like other natural processes, belonged and should be "restored" to its rightful role. The preferred means was lightning fire. Several parks were large enough and sufficiently well documented to conduct genuine experiments. They tested the idea that the fire practices of the native peoples were not markedly different from natural processes, that the real shock to these purely natural systems had come from the onslaught of fire suppression. But how, exactly, would lightning fire behave? How would its regime differ from what existed at the time of policy reform, and from what existed at the time of American settlement? What, in fact, was really being "restored"?

These experiments in wholesale fire introduction were subject to all kinds of distortions—the artificial fuels they inherited, the contrived zones within which they could burn, the general reluctance to tolerate high-intensity fire regimes, the stubborn tendency of smoke to drift outside borders. Not least is the brief period of time within which the records have accumulated. Even 25 years is but a blip for those regimes structured on the order of three or four centuries. "Restoration" itself proved a flawed ideal, drenched with irony. Still, fires occurred. They did so as events, however, not as experiments. The scale of the landscape was too vast and varied, the factors too slippery to track with experimental rigor. In any case, the outcomes are suggestive.

They suggest that, indeed, lightning fire regimes are varied: some sustain frequent fire, some fitful. They confirm that a few big years rack up most of the burning. Those years boast drought-blasted landscapes, lightning busts, long-burning fires that creep and sweep as conditions warrant. They suggest further that, over time, recurring fires arrange fuels into a mosaic, sometimes coarse, sometimes fine-grained. Although there are overlaps and gaps—a few sites burn over and again, others almost never—a kind of rough jostling keeps each part in it's place.

All this happens against a dynamic backdrop in which human fire practices, both the starting and stopping of fires, are yielding to lightning and "natural" fuels. A Second Fire regime is dissolving and a First Fire regime is congealing out of the mixed sludge that remains. Not least, the experiments suggest that the dewy landscapes explorers first witnessed were not solely or even largely the product of nature alone. Earlier peoples had shaped them, and most particularly had tended the biota-sculpting flames. The bold experiments are restoring fire; they are not necessarily restoring historic fire regimes.

Unsettling Fires

The Yellowstone epiphany. America's National Park Service reformed its fire policy in 1967–68 with the intention of getting more fire into the lands under its administration, particularly naturally caused fires. One outcome was the "prescribed natural fire," which allowed a lightning fire to burn if it did so under an approved set of conditions called a prescription. A fire could thus be both wild and controlled. Accordingly, Yellowstone National Park proposed a new program in 1972 that allowed natural fires to run their course over large segments of the park. In 1985, after some experience had accrued, the Park Service sought to bring Yellowstone's program into closer conformity with those of other parks. Yellowstone's fire plan was revised, but the park refused to incorporate into it binding prescriptions, which were of course the heart of the policy. The plan was still not officially approved (or even consistently applied) when major fires struck in the sumer of 1988.*

Lightly or severely, the fires burned off approximately 45 percent of the park. A media firestorm resulted. The park expended over $130 million in suppression costs. The American public received a crash course in the theory and ideology of prescribed natural fires. That a fire plan without prescriptions was, in fact, a let-burn (not a prescribed fire) program was lost in the furor. But the conflagrations did raise interesting questions about "natural" fire and its place on Earth. Probably no event in the 20th century alerted a larger audience to the ecological significance of free-burning fire. The park celebrated the fires as a magnificent "restoration" of Nature.

But a more nuanced interpretation is possible. Yelllowstone clearly had a fire deficit. Less land had burned over the past hundred years than over

*For a scientifically based summary of the fires, see the special issue of *BioScience* 39(10) entitled "Fire Impact on Yellowstone" (November 1989). A general guide to the literature is available in D. Despain et al., *A Bibliography and Directory of the Yellowstone Fires of 1988* (n.d.).

the centuries prior. The park attributed this fact to fire suppression, begun in 1886, although it also claimed that suppression had been ineffective until aerial fire control arrived with slurry bombers and smokejumpers in the late 1950s. Regardless, the summer's fires apparently stripped away an area of old-growth forest equivalent to what would likely have burned on average over the course of a hundred years. More puzzling was the absence of fire on the northern winter range. Fire-scarred trees along the perimeter of the Lamar Valley showed a minimal fire return of 35 years or so; probably a good chunk had known fire almost annually. Yet the 1988 fires—the largest on record—failed to burn them. Were the ecologies of forest and steppe so out of step? Or were the "missing" fires not those set by lightning and suppressed, but those that had over thousands of years been set by humans and were no longer allowed? Those fires had likely been thick as mushrooms—fires kindled to drive animals, prune berries, and scour openings; signal fires, camp fires, smudge fires that typically litter aboriginal landscapes and that can, during times of drought, romp over large landscapes. Was Yellowstone's fire deficit the outcome of suppressed First Fire, or the result of abolishing Second Fire? At present, there is no hard evidence to say conclusively, but analogies suggest powerfully that it was the latter. This means that the regimes that the park was "preserving" were not wholly natural and that a long era of adjustment was under way as lightning began to move into a landscape from which anthropogenic fire had been evicted.

The great fires also force us to consider the meaning of fire ecology. To what extent must even natural reserves include human behavior? A total of 31 fires hammered the Greater Yellowstone Area that summer. Outside of the park, every public agency recognized the serious conditions that prevailed and with two exceptions determined to fight (or try to fight) those fires from the moment of their first report. Until ordered to stop in mid-July, Yellowstone, however, accepted every start, considering each new fire, whether originating within the park or not, as natural. This clearly reflects institutional—that is, social—values, not environmental conditions. Once the fires got large they became uncontrollable. Effective containment was only possible at the time of ignition, and that required a willingness to immediately declare them wild and take action.

The other lesson is that human institutions—government agencies, media reports, scientific journals—can have weightier offsite impacts than ash washed into streams or smoke wafting through towns. They can hugely influence fire effects. The ecological consequences of the

Yellowstone fires did not remain in the Greater Yellowstone Area. They prompted a national review of fire plans by the federal land agencies, which shut down fire programs throughout the country, the strongest only temporarily, the rest more or less indefinitely. The Yellowstone fires were thus felt in Florida, Minnesota, Oregon, and New Mexico. They shaped how national park administrators in Australia, Canada, South Africa, the Soviet Union, almost everywhere thought about the role of fire and how they sought to apply or withhold it. The ecological outcomes were, in fact, global. Print media and television carried ideas and images around the world more thoroughly than convective winds lofted embers miles ahead of the flames or the swirling currents of the atmosphere picked up the carbon dioxide blown free by the combustion of lodgepole pine needles. It was far from clear where the wild ended and the cultural began, or that the distinction was one nature even sought.

Crossing the Threshholds at Kruger National Park. At two million hectares, South Africa's Kruger National Park is one of the great nature reserves in the world. But like so many others, it resulted from historical accidents that stripped away or culled much of the resident humans. By the beginning of the 20th century, long decades of intertribal wars, the Mfecane, and the final convulsions of the slave trade had hollowed out the landscape. Then came a decade of wrenching drought, the rinderpest epidemic, and reckless hunting that gutted native fauna and livestock, and Europe's notorious scramble for Africa that redrew the continent's political boundaries. Africa's fabled wildlife poured into this emptied Eden and filled it with the megafauna marvels that had survived the Pleistocene-closing extinctions. In 1898 the Transvaal established the Sabi Game Reserve, and in 1926 it became Kruger National Park.*

The lowveld is a place prone to burning. Probably the collapse of the fauna, savaged by rinderpest, had bumped the fuel loads higher, and thus stoked, for a while, hotter fires. In the early years of the reserve the fires simply happened because there was insufficient power to stop them. But Colonel James Stevenson-Hamilton, for 50 years the park's warden,

*For a good discussion of the issues, see W. S. W. Trollope et al., "A Structured vs. a Wilderness Approach to Burning in the Kruger National Park in South Africa," *Fifth International Rangeland Congress 1995*, pp. 574–575; B. W. Van Wilgen, H. C. Biggs, and A. L. F. Potgieter, "Fire Management and Research in the Kruger National Park, with Suggestions on the Detection of Thresholds of Potential Concern," *Koedoe* 41(1) (1998): 69–87.

recognized that "in a sanctuary for wild life" it was "essential to burn the old long grass, but this must be done methodically."* So it was, save for a few short-lived (and failed) experiments in full suppression. In 1954 the park was divided into 400 burning blocks, a third of which were burned annually after 50 mm of rain had fallen. Additionally, Kruger established what have become the oldest study plots for fire ecology anywhere.

Then philosophy and ecological science intervened. Both argued for a greater variety of burning. In 1975, park officials modified their practices to include a wider range of seasons and savannas. Further modifications appeared in 1980. Then in 1990 the old grid of burn blocks was scrapped in favor of 88 larger, more "natural" units. Three models of burning materialized. One burned patches more or less randomly, or as conditions permitted. Another modified the old deliberate burning to meet more precise ecological conditions, to keep the biota within certain threshholds. And in 1994 the park completed the trilogy by adopting for a large fraction of its holdings a Yellowstone-like policy of "natural regulation" in which anthropogenic fires were, where possible, suppressed and lightning fires allowed to burn. Two years later Kruger achieved a Yellowstone-like conflagration in which dry lightning burned a fifth of the park.

It is difficult to know what the natural conditions of Kruger might be, or how long it might take to "restore" them. Fire-tending hominids had occupied the landscape for over a million years; fire-starting hominids, for more than a hundred millennia. The place had never been without anthropogenic fire through all that time, save for blips like that which allowed the land to be first reserved. Yet there is no evidence that people burned with the same regime as lightning. What, then, was natural, and what the outcome of human history? A fire regime based solely on lightning would eventually establish itself; most likely it would be different from any that had ever existed. What this might have to do with the preserved landscape was unclear. As an ideal, it had meaning in testifying to the transcendence of nature. As a practice, it was often confused, ironic, hapless. As with Africa's politics, so with its ecology: decolonization had liberated, unsettled, challenged, and muddled. Its fires illuminate that confusion.

*Col. James Stevenson-Hamilton, *South African Eden: From Sabi Game Reserve to Kruger National Park* (London, 1937).

Chapter Three

Aboriginal Fire

CONTROLLING THE SPARK

Whether they nudged or hammered, subtly shifted a landscape or shattered one, first-contact fires had this much in common: they could not repeat themselves endlessly. Like all pioneers, they destroyed the conditions on which they depended. First-contact fire plunged biotas into an anthropogenic forge where, assisted by the sledges and tongs of hunting, gathering, herding, and farming, humans reshaped the land to their own purposes. A place that lacked fire now had it, and a place that possessed fire now had it in different forms. Neither could afford to let fire lapse without consequences. But how, exactly, could anthropogenic fire continue?

Often it thrived within an aboriginal economy whose landscape jugglings stopped just shy of outright agriculture. From afar, particularly as viewed by early 21st-century urbanites, aboriginal fire often appears quaint or perverse, like snare traps or a kind of ecological graffiti. To argue that small bands of humans without metal axes, servant livestock, and moldboard plows, much less bulldozers, could wrench whole biotas into new figurations seems fantastic. And to imagine that people would plaster fire on the landscape, or that they could derive much benefit if they did, mocks modern belief. In fact, their lives were often impossible without free access to fire. Unburnable landscapes were generally unlivable ones. Consider, too, that present-day tribes, driven into stony deserts and soggy rainforests, are poor models for those peoples who in the past inhabited more robust environments. Remember, too, that fire is extraordinarily interactive. It merges with other practices, with almost all that people do. It can penetrate into the very woof and warp of ecosystems. And not least it can rove far and wide; a single spark, properly timed, can rush over thousands of acres. So, too, aboriginal burning can—rightly positioned—sweep across continents.

Certainly their fires mattered deeply to the peoples who tended them. The more primitive their technology, the greater their dependence on fire. Alfred Radcliffe-Brown's portrayal in 1948 of the Andaman Islanders still stands as an eloquent testimony. Fire, he concluded:

may be said to be the one object on which the society most of all depends for its well-being. It provides warmth on cold nights; it is the means whereby they prepare their food, for they eat nothing raw save a few fruits; it is a possession that has to be constantly guarded, for they have no means of producing it [not true, but rarely exercised], and must therefore take care to keep it always alight; it is the first thing they think of carrying with them when they go on a journey by land or sea; it is the centre around which the social life moves, the family hearth being the centre of the family life, while the communal cooking place is the centre around which the men often gather after the day's hunting is over. To the mind of the Andaman Islander, therefore, the social life of which his own life is a fragment, the social well-being which is the source of his own happiness, depend upon the possession of fire, without which the society could not exist. In this way it comes about that his dependence on society appears in his consciousness as a sense of dependence upon fire and a belief that it possesses power to protect him from dangers of all kinds.

The belief in the protective power of fire is very strong. A man would never move even a few yards out of camp at night without a firestick. More than any other object fire is believed to keep away the spirits that cause disease and death.*

Where circumstances were more favorable, fire became far more than a symbol or a social hearth around which the band gathered. It became for people a means to mold the environment in their own image, in ways both huge and delicate. There were few aboriginal landscapes not smelted, seared, or smoked by the aboriginal torch.

Why They Burned

So aboriginal peoples burned; they had to, and they wanted to. The firestick extended humanity's reach far beyond its grasp. Even seafaring Tlingit in cold-temperate rainforests fired berry patches, as did Inuit atop frozen Arctic tundra. The issue is not whether aboriginal peoples burned, but why they burned and what the ecological outcomes were.

Fire's purposes are at once universal and particular. A forester in British India who surveyed the Ghumsur Forest in the early 20th century noted that all the state forests were subject to fires crossing from the numerous surrounding zamindari forests. "The latter," he wrote, "if

*Alfred Radcliffe-Brown, *The Andaman Islanders* (New York: Free Press, 1948), p. 258.

they are in a condition to burn, are always burnt.... Then in the large hill forests frequented by the Khonds the jungle is fired as a matter of course to facilitate tracking and for other well-known objects." This was mostly associated with foraging and hunting. The Khonds, for example, would not enter a tiger-ridden "jungle" without first burning it. He continued:

> In the lower hills and more accessible country bamboo cutters and permit-holders generally are responsible for a great deal of the mischief. Wherever a hill is frequented for bamboos there are always constant fires.
>
> Other causes are the practice of smoking out bees for honey—a very common origin of fire—of burning under mango and mohwa trees to clear a floor for the falling fruit and flowers; the roasting of Bauhinia seed; the burning of under-growth round villages and cultivations which might harbour tigers and panthers—this will probably prove one of our most serious obstacles to restocking the sal forests; and the spread of fire from banjar lands under clearance from cultivation....
>
> The long list of causes is almost complete if to the above are added the burning of forest by graziers, and for driving out game or finding a wounded animal.*

Add into this bubbling stew the unbounded use of smokey fires to drive off noxious insects, both those in the fields and those in thatched roofs. Add, too, the role of fire for general ecological cleansing, the mark of biotic housekeeping, and for laying claim, the sign of human possession. Those who burned a land asserted their rights of use. A properly burned land was the emblem of human stewardship. And add, finally, the power of sheer fire littering. Accidental and careless fires were strewn along routes of travel. Embers fell like the husks of opened nuts.

To return to Ghumsur Forest, the presence of livestock and nearby cultivated fields further seasoned the aboriginal mix of fire practices (India seems always to add to, never subtract from, its ecological pot), but the remainder of the litany could apply from Finland to Tierra del Fuego. Tennessee tribes burned to assist the harvest of chestnuts, Mesolithic Europeans for hazel and olives, Californians for acorns. In East Africa smoking out bees with torches, which then fall to the ground, has long been a major source of veld burning.

*S. Cox, in A. A. F. Minchin, "Working Plan for the Ghumsur Forests, Ganjam District" (Madras, 1921).

To Drive and to Hold: The Renewable Logic of Fire Hunting

The basic premise is: whatever might be hunted by fire *was* hunted. Fire drives are recorded for every conceivable game animal, from elephant to antelope, wallaby to rhea, to deer, moose, bison, alligators, woodrats, rabbits, and even grasshoppers. What wasn't driven by fire was attracted by the green-up of old burns. The habitats of bobwhite quail and Scottish grouse, the movement of springbok and wildebeest, the nesting sites of waterfowl and muskrat—all could be tweaked and plied by selective burning. Virtually all were.

North America abounds in examples. In early 17th-century Virginia, Captain John Smith reported how the Indians ("commonly two or three hundred together") could fire-drive deer within hunting grounds or off peninsulas where they would be easily slaughtered from canoes. In the early 18th century, John Lawson described the process in the Carolinas, where the Indians "commonly go out in great Number, and oftentimes a great many Days Journey from home, beginning at the coming in of Winter," and again "they go and fire the Woods for many Miles, and drive the Deer and other Game into small Necks of Land and Isthmus's, where they kill and destroy what they please." Cabeza de Vaca described similar practices in early 16th-century Texas: "Those from further inland ... go about with a firebrand, setting fire to the plains and timber so as to drive off the mosquitoes, and also to get lizards and similar things which they eat, to come out of the soil. In the same manner they kill deer, encircling them with fires, and they do it also to deprive the animals of pasture, compelling them to go for food where the Indians want." Lewis and Clark found it necessary to sink their canoes in the Missouri in part to avoid the prospect of their being burned by prairie fires. They added another variant of the fire hunt: "Every spring the plains are set on fire and the buffalo are tempted to cross the river in search of the fresh grass which immediately succeeds the burning." In the process they were often isolated on ice floes, floated down the river, and dispatched with ease by Indian hunters waiting at convenient sites. In Spanish California, José Longinos Martinez noted how the Indians had the custom of burning the brush, "for two purposes: one, for hunting rabbits and hares (because they burn the brush for hunting); second, so that with the first light rain or dew the shoots will come up which they call *pelillo* (little hair) and upon which they feed like cattle when the weather does not permit them to seek other food."*

In the early 19th century, Thomas Jefferson answered an inquiry from John Adams as to whether

> the usage of hunting in circles has ever been known among any of our tribes of Indians? It has been practiced by them all; and is to this day, by those still remote from the settlements of whites. But their numbers not enabling them, like Genghis Khan's seven hundred thousand, to form themselves into circles of an hundred miles diameter, they make their circle by firing the leaves fallen on the ground, which gradually forcing animals to the center, they there slaughter them with arrows, darts, and other missiles. This is called fire hunting, and has been practiced within this State within my time, by the white inhabitants.

Jefferson shrewdly suggested that this practice was "the most probable cause of the origin and extension of the vast prairies in the western country." So it probably was, not only in North America but wherever weather, terrain, and a grass-laden biota permitted broadcast burning by humans. And so, too, explorers recorded similar fire practices in the Sudan, Patagonia, Java, Guam, the Ivory Coast, and anywhere else hunters could induce fire.†

Plowing and Sowing with Flame: Firestick Farming

But aboriginal landscapes involved more than hunting. People fished, foraged, gathered, erected and decamped sites, and rearranged the biotic furniture of their ecological household to better suit their needs. Rhys Jones has coined the expression "firestick farming" to underscore the fact that Australian Aborigines, at least, did not passively yield to the landscape and let their fires merge seamlessly with nature's but actively

*John Lankford, ed., *Captain John Smith's America* (New York: Harper and Row, 1967), p. 22; John Lawson, *A New Voyage to Carolina* . . . , ed. Hugh T. Lefler (Chapel Hill: University of North Carolina Press, 1967, reprint of 1709 edition), p. 215; Adolf Bandelier, ed., *The Journey of Alvar Nuñez Cabeza de Vaca . . . 1528–1536*, trans. Fanny Bandelier (New York: AMS Press, 1973, reprint of 1905 edition), pp. 92–93; Lewis and Clark, quoted in Walter Hough, *Fire as an Agent in Human Culture*, Bulletin 139, U.S. National Museum (Washington: Government Printing Office, 1926), pp. 62–63; Martinez quoted by L. J. Bean and H. W. Lawton, in H. T. Lewis, *Patterns of Indian Burning in California: Ecology and Ethnohistory* (Ramona, Calif.: Ballena Press, 1974), p. xix.

†Jefferson to Adams, May 27, 1813, quoted in "Thomas Jefferson on Forest Fires," *Fire Control Notes* 13 (April 1952): 31.

intervened and burned to prod and push the biota into forms they found more desirable.*

There is no reason to believe that such experiences were limited to Australia. The kinds of fire practices chronicled by observers are remarkably universal. Patch burning to promote the growth of berries, fruits, or flowers is virtually identical in British Columbia, Maine, India's Madhya Pradesh, and New Zealand's North Island. An 1887 description of California could be repeated in Mozambique, Brazil, and Greece: "In the spring ... the old squaws began to look about for the little dry spots of headland or sunny valley, and as fast as dry spots appeared, they would be burned. In this way the fire was always the servant, never the master.... By this means, the Indians always kept their forests open, pure and fruitful, and conflagrations were unknown."†

The open landscape was often a more desirable one. On it ungulates could browse and graze; across it hunters and warriors would watch for prey or raiding parties; and over it tribes could travel untrammeled. A landscape regularly cropped by fire, moreover, bore the unmistakable trace of the human hand. Keeping woods free of underbrush, burning back scrub, and quelling fuels were all means of cleaning up the countryside, or exercising the rights and duties of biotic citizenship. Of course, the reverse of those socially prescribed fires that sweep and polish the landscape are those set by the wanton, the reckless, and the malicious. Nuisance fires swarmed around human groups like flies. To see them was to know that people were present.

Where and How They Burned

Pulses and Patches

Nature supplied the rough matter, the inspiration, and the models. Nature furnished flame and fuel; nature's fire regimes sketched the rough rhythms for burning; nature's quilt, the coarse patchwork of fuels. But aboriginal peoples captured nature's fires and redirected them. They seized the landscape mosaic they inherited and fashioned a new regime by changing fire's timing, its scale, its frequency, its intensity. Within the limits imposed by their toolkits and geography, they replaced nature's

*Rhys Jones, "Fire Stick Farming," *Australian Natural History* 16 (1969): 224–228.

†Joaquin Miller, quoted in Harold H. Biswell, *Prescribed Burning in California Wildlands Vegetation Management* (Berkeley: University of California Press, 1989), p. 48

work with their own. In the words of Henry Lewis, they substituted fires of choice for fires of chance.

While the landscape remained one of pulses and patches, it felt the hand (and the mind behind the hand) of humanity. Social rhythms compounded, and sometimes rivaled, those of weather. Human chipping and trimming, over a slow fire, helped size and shape the biotic pieces. The direct effects of applying and denying fire yielded big changes. But even larger were the often-leveraged indirect effects that resulted because aboriginal peoples also hunted, sometimes to extinction; they transferred and harvested plants, selectively and widely; they migrated with the seasons, adding and removing fire well outside the beats of the local climate.

Yet the fundamental logic was simple: burn early, burn light, burn often. As various scraps of the landscape dried sufficiently, they were fired. As the dry season progressed, the patches would grow accordingly, but the larger landscape would be crossed with the traffic of burned corridors and dappled with green and black patches. It was essential to protect critical or sensitive habitats, not only villages but sites that produced fruits or useful cover that could not survive regular firing or high-intensity wildfire. By the time fire season had deepened, such places were insulated by protective swaths of early-burned fuels. Only through controlled fire could they be spared from wildfire.

Aboriginal peoples burned whatever the land would bear. If fuel existed, fire followed. Often aborigines carried their firesticks with them, constantly dribbling embers and scattering sparks and kindling tussocks and shrubs and hollowed-out trees wherever they went. When they stopped, they lit fires. When they traveled, fires followed like camp dogs. When they wanted to extract some resource, fire was there as an enabling device. They foraged for fuel as they would for mushrooms or edible lizards. Such practices appear to be universal. Anthropogenic fire goes where people go, and nowhere more than with fire-toting aborigines was the reverse also true: people tend to go where fire is possible.

Lines of Fire, Fields of Fire

As aboriginal peoples cycle through landscapes, so does their burning. Fire lights their corridors of travel, and it sweeps the plots where they hunt, harvest, and camp. It both scrubs up the countryside and litters the landscape with open flame. Since hunting and foraging peoples tend to migrate through their territory, tapping sundry foodstuffs according to a calendar of seasonal availability, their fires follow that

FIGURE 4. Lines of fire. People deliberately burn (and fire-litter) along routes of travel. The photo on the left shows a major corridor through the Black Hills as it existed in 1874. The ridge on the left faces south, that on the right, north; thus they support very different forests and fire regimes. But both flanks exhibit plenty of evidence of burning.

The graph derives from fire-scarred trees in the Northern Rockies. Yet there is good reason to believe that the scenes it records developed under a very similar regimen of forests and fire. Read the graph this way: the smaller the bar, the more frequently fire returns to a site. For each of three historic periods, the sampling records the fire frequency on places known to be regularly inhabited (white bar) and for places remote (stippled bar). Before settlement by Americans, there is a clear correlation between the abundance of fire and the frequency of human use. The contact period began to scramble this segregation as miners, in particular, sprawled over the landscape. In the 20th century, with the land largely protected as reserved forest, the old distinction ends and fires disappear. The ancient landscape mosaic vanishes as well. The photo on the right shows the outcome a century later. The woods have become more uniform and the potential for intense fire greater. (Sources: Barrett 1980 [graph data], redrawn by the University of Wisconsin Cartographic Lab, and Progulske 1974 [photos reproduced by permission]) same aboriginal almanac. The hearth fire spreads beyond windbreak and hut to embrace whole landscapes.

In a rough way, the volume of human traffic determines the volume of burning, but because one person can ignite an unbounded number of fires and because every one of those fires has the capacity to race far beyond its point of origin, the density of humans does not by itself dictate the size of the area burned. Not ignition but fuels decide whether a start sprawls across the countryside or dies out. Fire must interact with the surrounding vegetation, only a portion of which is available for burning at any given time. Scrutinizing Australian Aborigines, however, Rhys Jones reckoned that a single wandering band could ignite 5,000 fires a year—an estimate he regarded as very conservative; this on the hottest and driest of the Earth's continents.*

Collectively these lines of fire and fields of fire stitch together a new landscape quilt within which natural fire, if it occurs, must operate. Typically, the burning begins early, as soon as fuels can accept it, and continues throughout the dry season. By the time lightning fire arrives, large fractions of the landscape are already scorched or otherwise unavailable. Lightning-ignited fires can only feed on the unburned sites. Whether these are big or small depends on the larger geographic features of the landscape; how mountainous it is, how droughty, how wet, how plumped with fine fuels. In this way—and in virtually every environment—humans sew the patches and pulses into fire regimes that meet their own ends.

---

*Rhys Jones, "Fire Stick Farming."

FIGURE 5. Competing geographies of fire. Not only do people compete with natural fire by season, they compete by place as well. These two maps illustrate this process nicely.

Alaska shows the outcome with particular vigor because it has legally restricted, and hence separated, human use from much of the landscape. The result is that anthropogenic fires cluster around villages, cities, and modern routes of travel (top map); lightning fire sprinkles the interior as moisture surges through the great valley of the Yukon during the summer (bottom map). Remove those legal proscriptions, and the anthropogenic fire regimes would compete directly with those of nature. Until the practice began in the 19th century of reserving large swaths of land from permanent habitation, direct competition was the normal pattern. Particularly where surface burning is easy because of grassy cover (as it is not in Alaska), the geography of anthropogenic fire dominates the overall geography of fire. (Source: Gabriel and Tande 1983, redrawn by the University of Wisconsin Cartographic Lab)

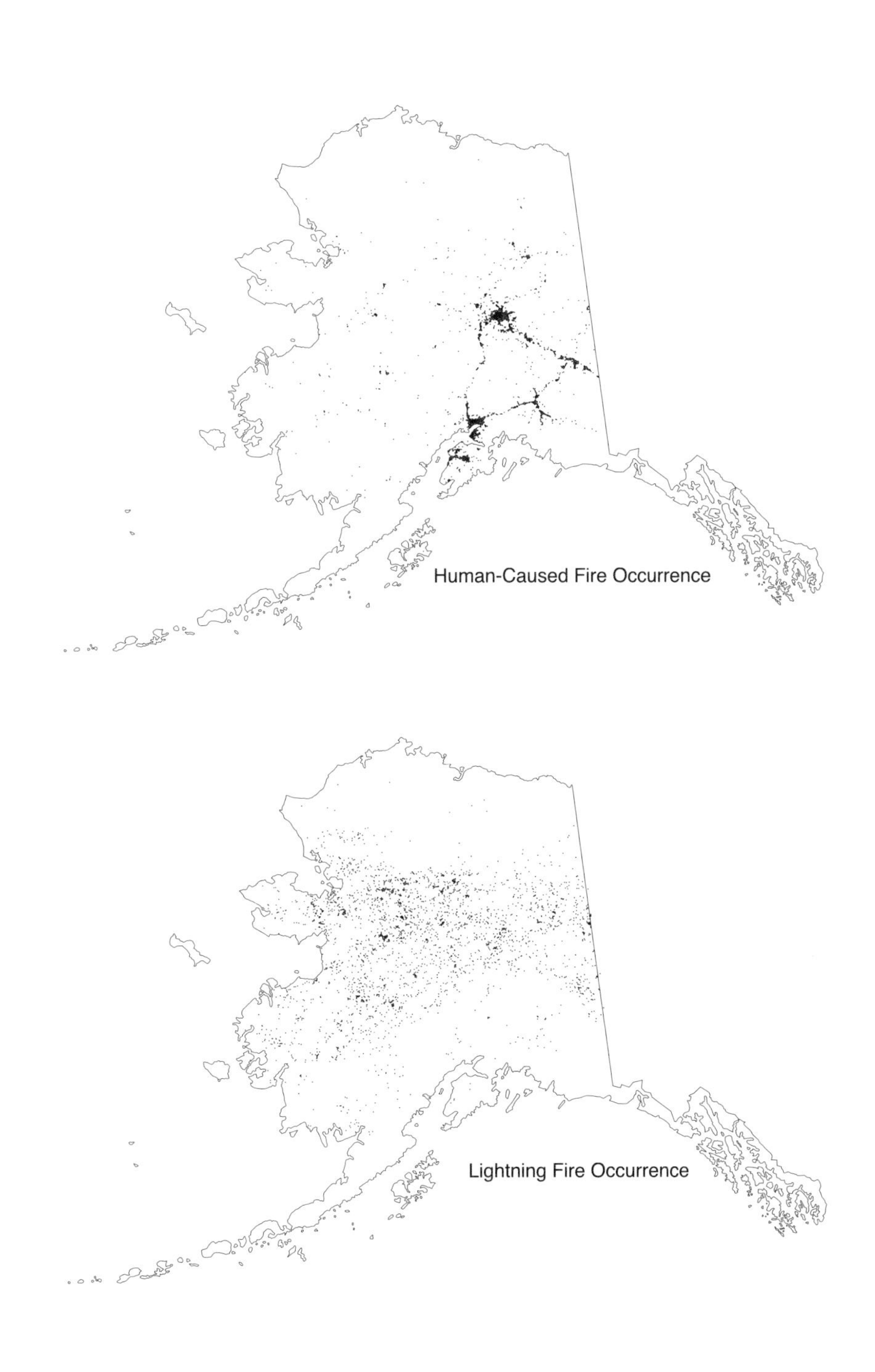
Human-Caused Fire Occurrence
Lightning Fire Occurrence

This is never a rigid or mechanical system. Fires remain organic, opportunistic, full of the randomness and even whimsy of human life. Always there are quirks, and every year is exceptional. No cycle repeats itself with the fixed logic of a piston returning with each revolution to top dead center. Yet patterns—fire regimes—do emerge. Even with a constant human flame, some places know fire annually, others only on the order of decades or centuries. The firestick was only as powerful as what it touched. To achieve a tighter grip over regimes required control over terrain, weather, and fuels. Yet, whether they grasped the land by the throat or led it by the hand, aboriginal peoples took it in new directions and ultimately branded it with their own character.

Fuelbreaks, Firebreaks

The aboriginal spark sometimes took and sometimes didn't. It was most effective when aborigines moved into a land already flush with fire or one capable of burning but lacking regular ignition. Aboriginal burning proved most successful where it could widen fire-wedged cracks that already existed; where it could maintain a fire-driven landscape rather than create one from scratch; where the hunting of megafaunal herbivores could free up more fuels; where fire-hungry brush, grasses, and weedy scrub-saturated biotas waited for fire to unfetter them; and where climatic wobbles, especially strong swells of wet and dry, were common.

Much of the Earth offered just such conditions. European explorers repeatedly encountered great savannas and prairies—"champion fields," the British called them—that reminded them of the pastoral landscapes of England and France. And in truth most were as much the outcome of human tampering as were the scenes the explorers referenced. In the absence of livestock, wild fauna hunted by fire had served the same purpose. This was no untouched wilderness: it was a made place. And it was a place often made possible only with fire.

Still, much of the Earth proved more unfriendly. Aboriginal burning was only as powerful as the amount of fuel available to it—combustibles of sufficient kind and amount that they could carry fire at least seasonally. Aborigines' inability to create new fuels, not their capacity to kindle fire or their willingness to use it, was what hobbled their burning. Aboriginal societies could rearrange fuels by hunting (and of course by burning), and for some lands this was enough. Many lands, however, proved tough to crack and, more ominously, not a few turned hostile to fire. These were lands awash with wet, shaded woods; lands that dried only rarely; lands that had few giant megafauna that pushed over, uprooted, or browsed away canopies and exposed understory to sunlight and wind.

Such lands resisted fire colonizing. Firesticks bounced off them like stones thrown against a granite cliff. Without some point of entry—the soft pounding of a wet-dry cycle, particularly—fire could not wedge open such stubborn biotas, or if it flared, it could not rush boldly outward. More poignantly, lands that once boosted fire might come to repel it. Perhaps they might shed their seasonality, lose their great browsers, replace flame-hungry combustibles with sodden lignin. If that happened, those aborigines who stayed were left to huddle around their windbreak-shielded flames and turn to rivers, lakes, and seas for sustenance. For most aboriginal cultures, the dying fire rightly symbolized a dying people. It could as well stand for a dying landscape.

Dying Fire: When The Firestick Leaves

The dying fire can speak as trenchantly as the living one. Some of the most profound ecological measures of anthropogenic fire have come from observing the consequences of removing it, for the fire regimes that aboriginal peoples laid down were as fundamental to landscapes as the rhythm of the rains and the cycle of green-up and curing. Whether anthropogenic fire was vigorously promoted or just tolerated by the biota matters little: it simply *was.* And it *remained,* often for centuries, more than long enough for the biota to adapt to its alchemical heat, flame, and smoke.

Yet few aboriginal landscapes have survived. Most passed into agriculture. Many slid almost seamlessly into fields, patch by patch, year by year, as tame browsers replaced wild ones, row crops replaced wild forbs and tubers, and fires for farming and herding replaced those of foraging and hunting. But displacement could also be sudden. Farmers invading closed forests could obliterate an aboriginal landscape. So could pastoralists swarming over arid grasslands. In such instances the old landscape might vanish almost overnight, its woods burned or its grasses gobbled and crushed, unable to burn because its ready fuels had been stripped away. Yet it also happens that aboriginal fire can vanish without leaving an ecological heir. The land may be abandoned for a long time, or more often in recent decades, it might be converted into a nature preserve. What happens next will depend on the innate fire-proneness of the place. If lightning fire is possible, then fire will eventually return in some form. If not, not.

Fading Fires: How They Change Landscapes

Fire's removal upsets landscapes, and an abrupt end can shock them. A landscape that has known a particular fire regime for many times longer than the age of its oldest resident may suffer from fire's withdrawal. A biota used to winter frost can languish without it, as will a land used to long winter rains that receives only a summer downpour or two. Such changes can shake an ecosystem to its roots. Yet that, in brief, is what has often happened with many fire-dependent landscapes from which fire has fled. The disappearance of aboriginal fire can unravel a biota as fully as diseases can its demography. Without its tenders, anthropogenic fire, like its fuels, either goes wild or goes out.

Sydney savannas (Australia). The best observations come from colonizing Europeans, especially those from societies that had experienced the Enlightenment and had begun to industrialize. The contrast between peoples and places was too great, and too interesting, to ignore. Writing about Australia in 1848, Surveyor-General T. L. Mitchell remarked:

> Fire, grass, and kangaroos, and human inhabitants, seem all dependent on each other for existence in Australia; for any one of these being wanting, the others could no longer continue. Fire is necessary to burn the grass, and form those open forests, in which we find the large forest-kangaroo; the native applies that fire to the grass at certain seasons, in order that a young green crop may subsequently spring up, and so attract and enable him to kill or take the kangaroo with nets. In summer, the burning of long grass also discloses vermin, birds' nests, etc., on which the females and children, who chiefly burn the grass, feed.

When those fires disappeared from around Sydney, Mitchell noted the consequences: "kangaroos are no longer to be seen there; the grass is choked by underwood; neither are there natives to burn the grass."*

The burning was, as Mitchell remarked, a "simple process." The annual tracks of Aboriginal songlines became threads of fire that stitched together a quilt of burned patches. But the process was never mechanical,

*T. L. Mitchell, *Journal of an Expedition into the Interior of Tropical Australia* (London, 1848), pp. 412–413.

never identical year by year. Peoples could be sick, or at war, or careless, confused, or in a blue funk, or the seasonal rains might come too early, too late, or too often—all of which upset the burning. The general outcome, however, was to further grasses at the expense of dense woods. The practice maintained precisely those fine fuels which could carry fire and further project the most powerful of Aboriginal techologies. The extinction of those fires eliminated their fuels, and without fire the woody scrub overran the landscape.

With remarkable fidelity, the replacement of unburned grasslands by scrub has repeated across the globe: the leafy trees that overran tall-grass prairie in North America, the woodland and boreal forest that moved south across the previously grassy Canadian plains, the woody thronging onto Brazilian *cerrado* and *campo*, the chaparral and dog-hair thickets that overgrow California, the dense understory that carpets once-open eucalypt forests, the juniper woodlands galloping over the American West, the brush that has gripped the overgrazed and under-burned sourveld of southern Africa, the closing of open-patched landscapes everywhere—the list is endless. In all these cases aboriginal fire departed and no other regime of anthropogenic burning moved into the vacuum. Without those flushing fires the area becomes overgrown, much as canyons, deprived of annual floods, begin to choke with boulders and debris. Without controlled fires there would only be wildlfires.

Tallgrass prairie (USA). At the time of contact, the eastern, more humid Great Plains swelled with tallgrass prairie. In habitually wet sites its forbs and grasses lay hidden under woodlands, but freed of those shade-casting trees, they could dominate—dappling the landscape in splotches, thrusting eastward as vast prairie peninsulas and immense "barrens," and spreading into a sea of grasses over the heaving plains. The prairie burned regularly, once every three years on average; more often as weather and grazing allowed. On the open plains one fire could rush in long, twisted fronts for miles. In the more densely rivered east, the landscape fractured into smaller shards and slivers, each demanding its separate ignition, a density of fires that people alone could kindle.*

*The literature is immense and often site specific. For a general survey, see Scott L. Collins and Linda L. Wallace, eds., *Fire in North American Tallgrass Prairies* (Norman: University of Oklahoma Press, 1990), and for an inquiry into anthropogenic burning, K. F. Higgins, "Interpretations and Compendium of Historical Fire Accounts in the Northern Great Plains," *Resource Pub. 161* (Washington: U.S. Fish and Wildlife Service, 1986).

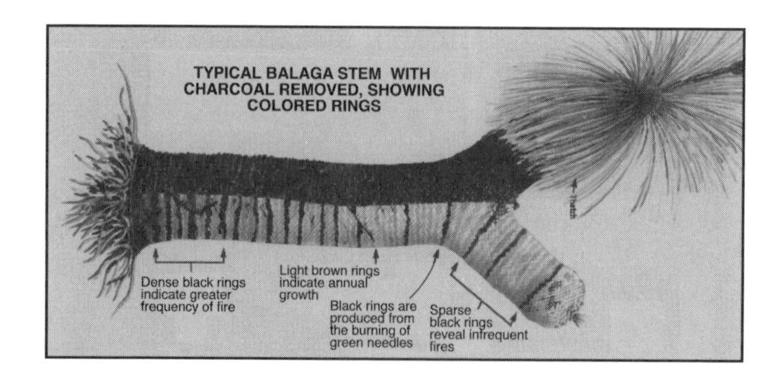

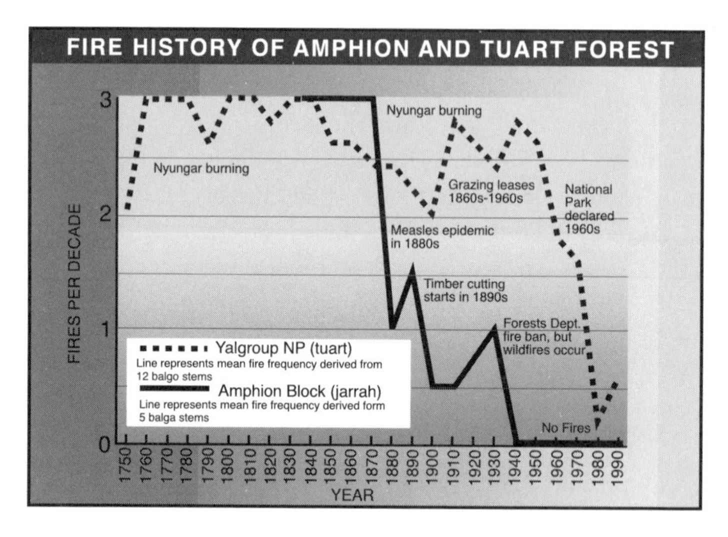

This was a landscape only anthropogenic burning could sustain. The climate and soils were fully capable of supporting forests; and when the fires moved out, brush and trees swiftly overran a site. Often the scrub encroached from the woody borders, where fires had formerly beaten them back, or they crept from fire-free refugia such as bottomlands or rocky outcrops. Studies suggest that pockets of trees survived within the oxbows of winding streams, but almost always on the eastern flank for the simple reason that the prevailing winds blew from the west, even during frontal passages. These winds drove free-burning fires eastward, until they struck streams or stony cliffs. The fires could thus clean out the windward side, but spared the lee, and it was here that, shielded from flame, fire-sensitive species survived. When the flames faded away for good, the woods slipped from their fire leash and roamed over the land.

FIGURE 6. Torch's end. Many Western Australian landscapes show fantastic adaptations to fire. The diagram shows the grasstree, which afflorsces after immolating in flame and whose stem scars in ways that can be dated (see picture). A close reading of that charred chronology reveals the power of Aboriginal fire. The graph records the outcome in two sites.

Prior to European contact, the Nyungar tribe burned the balga (as they called it) probably three times a decade. But prolonged contact broke that regime. After a massacre in 1834 and a measles epidemic in the early 1880s, the Nyungar around Dwellingup almost vanished (solid line). They took their fires with them. Some burning flared as timber companies strewed the land with slash in the 1890s. Then the forest department geared up for outright exclusion and more or less succeeded, save that infrequent large fires tended to replace frequent small ones. Near Yalgorup (dashed line) the regime began to decline by mid-century, then revived slightly as graziers, using Aboriginal labor (and fire practices), spread over the scene. Herds competed with fire for the grasses, but herders were often keen to burn for "green pick." A new regime resulted still relatively flush with burning. Then the pastoral leases expired and the Crown lands were declared a national park in the 1960s. At this point the fires all but disappear. The fantastic fire-adaptations that this biota exhibited had resulted from human fire practices. A landscape that had, for tens of thousands of years, known a particular pattern of fire suddenly had to cope with another. Possibly lightning would pick up some of the slack, but this is not a climate prone to thunderstorms. People had put fire into the land; people would have to again. (Source: Ward and Sneeuwjagt 1999, picture reproduced by permission, graph redrawn by the University of Wisconsin Cartographic Lab)

For the most part these patches, often large, were fired for hunting, often for bison. Historical accounts suggest that northern tribes of the Great Plains burned outlying areas during an autumn hunt, which forced the herds to seek out unburned patches for forage. These herds now resided near encampments, which made hunting over the long winter easier. During the spring the fall-burned patches greened up and drew herds back to the outlands, while the winter sites could be burned and made ready for the fall.

Inevitably, the process featured more complexity than this. Fires escaped; storms, drought, and winds upset the timing and scale of burning; accidents, enemies, and lightning all fired range at inconvenient times and places; and the wildlife itself was a cause as well as an effect, since it competed with fire for biomass. Once-burned patches, heavily grazed, might lack sufficient fuel to carry a hot fire, and so, unburned, might fail to sprout the tastier grasses and forbs for the coming season. Less intensively grazed, these sites would burn better during the next round, green up more vigorously, and again tempt the herds back. Not least, the character of hunting helped determine the population of herbivores, and thus the quantity of fuels available for burning. But fire there was. Without it, the ecological machinery slowed or ground to a halt.

The mythology of American settlement has celebrated the ax, the saga of forest felling. But in the tallgrass biota, settlement *brought* woods. Indigenes, and their aboriginal firesticks, left; roads and plowed fields carved firebreaks across fuels; livestock cropped off the grasses; and what land was not converted to town or farm was isolated, severed from far-propagating flames and no longer fired in its own right. Soon, colonists reported the outcome. Brush and woods were burying the barrens, swallowing up unfarmed prairie peninsulas, and invading the plains. When it was founded, St. Louis sat atop a grassy bluff. Settlement extinguished the formative fires, however, and the maturing woodlands tracked the history of these departed fires. The closer to the town, the denser and older the trees.

The prairie has survived only incidentally. One of the largest swaths, the celebrated Konza prairie in the Flint Hills of Kansas, endured because the stony country could not be easily farmed and remained within a tradition of herding predicated on spring burns. Drovers even wrote the dates for burning into contracts. Fire endured, and with it, the prairie. Places that have sought to reconstruct true prairies have found they have to restore fire to do it. By itself fire cannot conjure prairie from a wreckage of weeds and gulleys; fire exclusion did not alone eliminate prairie, and fire's return will not, unaided, revitalize it. But without fire even ceremonial pockets of native tallgrass will not thrive, and may not survive at all.

Banff National Park (Canada). Banff tells a kindred story. Here the abolition of aboriginal fire has caused coniferous woods to thicken, the prairie to contract, and aspen groves to collapse, smearing a once dappled landscape into a common green gunk. The decline apparently began when aboriginal tribes left and took their torches and spears with them. Rather than merely cropping the valley's lush wildlife, snipping off the surplus, the aborigines had, as a keystone species, organized the whole system. They seem to have acted like a school of muskellunge plopped into a pond, immediately restructuring the pond's food chain. The change soon rippled through the Canadian Rockies' great trough valley.

How did this occur? Hunters had long but seasonally inhabited the greater Banff region. Judging from bone deposits, they fed mostly on bighorn sheep and a mix of bison, deer, small game, and a few elk. They burned the valleys regularly, and left campfires alight wherever they wandered, keeping fire on the landscape without regard for lightning's lottery. From time to time, drought and winds combined to finger those flames deeply into the surrounding forests and occasionally to send them sailing through the canopies. These were exceptional years, of course; but they are the ones that most mattered biologically. Meanwhile, the big game sagely avoided the hunters' camps. Burned yet protected from browsers, those sites—and those others scoured by episodic crown fires—blossomed into swaths of aspen.

As the natives fled, however, they carried their fires with them. For a while fires from pioneering newcomers and the spark-casting locomotives of the Canadian Pacific Railroad kept the landscape in some kind of flame. The creation of a national park, Canada's first, in 1885 imposed a different pattern of protection. Predators like the wolf were driven off; fires were suppressed; tourist hotels and a golf course replaced tent encampments and fire-flushed hunting grounds. No longer pursued by either humans or wolves, the elk population waxed, then exploded. No longer burned, the short-lived aspen ceased to rejuvenate old clones or repopulate declining sites. What suckered upward, elk cropped off. By the 1990s, Banff had many tourists and elk, and few fires and aspen.

The prospects for reform on any significant scale are poor. The town presents daunting hazards for large-scale burning; the elk (even the town's "punk elk" that skulk through its streets like youth gangs) are regarded as a tourist asset; and the existing landscape, being overgrown and green, has become the new norm. Reintroducing wolves has done little good. The wolves avoid human habitations, thus cramming the balky elk into a closer huddle with the town. Controlled burning has proved too sparing, for while it stimulates aspen suckering and seeding by the millions, the elk soon scour off the fresh growth. Proposals for large-scale firing alarm what has become an urban complex in the heart of the reserve, along with environmentalist critics, themselves mostly urban, who want no warrant for or evidence of the human hand on the land. They dismiss aboriginal practices as trivial. The aboriginal landscape—the biota for which the park had been established—is blowing away like old smoke.

Faded Fires: How They Challenge Ideas

So have aboriginal fires languished throughout the globe. In most places another regime of hominid fire has replaced them. But not everywhere. And from time to time, as at Banff, the fallout from the fires that haven't happened may prove more toxic than the soot from those that do. The failure is not solely one of practice, of a nature too ornery to accept human ideals. The crisis may reside, instead, in those ideals. What, after all, is a "natural" landscape?

It is easy to say what is not natural. Cities, refineries, plowed fields, and sown pastures—human tinkering has clearly shaped them all. It has proved tougher to recognize that "primitive" peoples may have influenced their habitats, that they preferred to mold the landscape rather than to be molded by it. For many years, those who have sought to protect natural landscapes and those who have championed native peoples have found common cause, and have collectively protested the industrial storms that have broken against the indigenous world. It is hard for preservationists, in particular, to appreciate the extent to which aboriginal societies of hunters and foragers can shape landscapes, the degree to which what they perceived as "wild" might be "cultural."

Yet the historical record suggests that humans have shaped every place where they have lived, and where fire is possible they have the capacity to redefine those places wholesale. Aboriginal peoples have invented landscapes, nourished landscapes, pruned and sculpted landscapes. In the right circumstances, one does not need large numbers of people to yield big effects, because fire multiplies their presence. Fire propagates, fire catalyzes, fire enables. Remove those effects and you remove the props that help hold a biota in place.

The passing of fire can jolt landscapes as much as the draining of Lake Bonneville affected the Great Basin or as Dutch elm disease reshaped North America's temperate forest. The removal of fire has consequences: this is as true for aboriginal fire as for natural fire. To restore original conditions it is, moreover, not enough to restore flame. What is needed is the return of the aboriginal fire *regime,* a particular pattern of fire foraging, hunting, cleansing, and littering. Lightning fire doesn't do this; nor does agricultural burning; nor the prescribed burning beloved by contemporary fire strategists. If one wants aboriginal landscapes, one needs aboriginal fire regimes.

Chapter Four

Agricultural Fire

CULTIVATING FUEL

Aboriginal fire struck the Earth like a Promethean spark. But that spark was, in the end, only as good as its combustible surroundings. The firestick was more limited by its length than by the brilliance of its flame. A goodly hunk of the Earth remained untouched by flame, or visited only on a cycle of centuries. That could change only when humans controlled fuel as they did spark. Which is, from a fire-historical perspective, what the arrival of agriculture meant.

It meant fuel: it meant that people could create—cultivate—the fuel fire craved. The axes and hoofs of agriculture broke open biotas that First Fire avoided and aboriginal fire could only curse. Equally important, the *idea* of agriculture set in motion a huge foraging for suitable combustibles. Choppers, plows, shovels, rakes, all could pry apart closed forests, thick scrub, tough sod, and deep peat, and let fire enter. So, too, herded goats, sheep, horses, swine, and cattle could shake and split woods and shrublands and steppes into kindling and cordwood. The aboriginal firesticks followed those fuels, and this led to the cultivated field.

What fire got, it also gave. Without fire, agriculture was mired in floodplains and potted into kitchen gardens. Farmers and herders could only expand through disturbing new lands, but in controlled ways. These often required fire. There was no point in slashing without the hope for burning; no chance to browse dense woods without fire to free up space, or to graze steppes intensively without fire to renew the forage; and no prospect of keeping the native plants (and imported weeds) from overgrowing the cleared fields without flames to help beat them back. So long as agriculture could chop or grow combustibles, people would burn. Whatever else they sowed or reaped, they had to cultivate fire first.

The Fire in Agriculture's Hearth

Little of agriculture lacked fire, and much of farming and herding did not work without it. But was the catalytic flame also in some way a creator? Scholars have long referred to the originating sites of domesticated plants and animals as "hearths." Is there a reality behind that metaphor?

Likely there is. The consensus centers of origin for cultigens and livestock are virtually all in fire-prone environments. For plants there are well-defined wet-dry seasons, and for animals, seasonally available pastures, typically along the flanks of mountains. Both circumstances are ideal for burning. These are places that provide examples of fire-sculpted terrains, ready for people to seize or copy; landscapes that humans can tinker with through fire; sites that hold species which humans can domesticate. Early cultivators selected for those species they liked and that did well in the disturbed sites, and then they took control of the act of disturbing that made it possible to sow plants where they did not naturally belong. Those cultigens were a kind of controlled weed that, with human tending, could seize a site fleetingly wiped clean by human burning.

What we call "agriculture" thus became a practice of selective substitution, first of species, then of landscapes. The firestick farming of aborigines selects plants from among those that already exist at a place. Ax-and-plow farming goes further and creates suitable habitats for plants that come from elsewhere. Ultimately, it may fashion whole ecosystems: the "farm" brings its own plants and animals, fixes their relationships, lays down pathways of energy and nutrients. Such a system can even be exported in defiance of climate. The European agricultural mix, for example, collected together cereals, pulses, and herbivores from the winter-rains eastern Mediterranean, then thrust them into the summer-rains regime of temperate Europe. Something had to jolt and jostle the land for it to accept so startling a change. That something, of course, was fire.

Nature supplied the model. Swidden farming mimics First Fire's storm-slashed and lightning-kindled woods. It is a small step, one many aboriginal peoples took, to assist the regrowth rising in the ash or to carry other plants to the cleaned site. The next step is to create those slashed and burned plots themselves. So, likewise, pastoralism echoes the movement of wildlife as they follow seasons and the patch-burns of green, scorched, and dormant forage. Replace the wild herds with domesticated livestock; then cycle them through similar landscapes; then create those landscapes by cutting, grazing, and burning.

In fire-flushed and disturbance-rich places—which most agricultural hearths were—this transition can occur piece by piece, with the domestic replacing the wild as one might replace the tiles in a mosaic. People only had to tweak fire regimes to better suit their purposes; the border between aboriginal foraging and agricultural harvesting is murky because their fire practices are almost genetically related. That frontier appears most stark and the encounter most shocking when agriculture crashes into new lands, particularly places in which fire is scarce. When sun-craving cereals and grass-munching sheep and cattle try to enter sun-starved or tranquil woods, they require a more violent wrenching. Firestick farming need only massage the environment; ax-and-plow farming requires the ability to force fire whether the landscape naturally accepts it or not. Domestication requires more than simply loosing the hearth fire into the bush, which is no better than daubing the scrub with a firestick unless fuels are there to accept it. That—the labor of creating combustibles—is, from a fire-historical perspective, what moves the practice into true agriculture.

Without suitable fuel, there could be no fire, and therefore little farming. If nature did not freely furnish those fuels, the agriculturalist would have to invent them. Curiously, those landscapes that had little natural fire offered better prospects for control because there would be no rivalry with lightning and fewer chances, due to weather, for fire to escape. Fire would exist only to the extent that people chose to put it there. The fire regime would be theirs not only because they alone controlled the spark but because they controlled the fuels, which as often as not they had to gather or grow for that very purpose. Ax and hoof acted as ecological fulcrums for the firestick, allowing it to pry open the toughest biotas and to shoulder aside whole landscapes.

How to Cultivate Fire

Field and Fire: Regimes of Fire-Fallow Farming

There are agricultural systems that do not require fire. Riverine agricultures, for example, work by accepting routine flooding in place of routine burning. Irrigation can extend these processes elsewhere such that cotton and melons can grow in deserts and wet-rice cultivation in Asia can panel whole hillsides with water-retaining terraces. Even so, fire often crackles outside the waters or when they recede. Typically the postharvest stubble is gathered as fuel for the hearth or it is fired where it stands; beyond the floodplains proper, farmers burn to unclog landscapes for hunting and foraging, but it is not obvious that fire is mandatory for this brand of agriculture to work. Instead, water purges, water promotes, water determines the regimen of planting.

For farming to leave the scenes of those scouring and enriching floods, however, or to push into places where irrigation is difficult, farmers had to burn, and to burn again and again. Fire prepared the fields, fire continually renewed them, fire helped set the rhythms of their cultivation. Fire-floods swamped the native flora and recharged the fields with ashy silt. Fire shocked a site such that, for a while, it could be stocked with exotic wheat, carrots, turnips, cattle, goats, and ragweed. Had farmers shunned fire, the imported cultigens and livestock would have had no advantage over native species. Had they removed fire, the fields might rapidly revert to waste and wild. The ecology of such agriculture was necessarily an ecology of fire.

Yet the regimen of cultivated burning was a compromise between the needs of fire and the needs of cultivation. Excepting some outliers—true wildlands, or the Nilotic, Mesopotamian, Yangtzean floodplains—fire and field came to share a mutual geography. Neither could leap beyond its natural hearth without the other. What linked them was fuel, or what agronomy termed "fallow." Agriculture tilled fuels to feed fire as surely as it did grain and pulses to feed oxen and people. Call it then what it is, a fire-fallow system.

How the Farm Behaves as a Fire Cycle

The cultivated field rotated, and fire helped turn the crank. Sometimes the field hopped through the landscape, shunting from site to site as fuels dried, crops matured, and weeds invaded. The larger scene was a controlled jumble of patches cut, burned, abandoned, and reslashed, refired, and released once again. Sometimes, however, when the farm was fixed by law or custom to a single locality, one scene succeeded another as soils wore down and new plants were sown to replace the old. Eventually the field fell to fallow, which burned and turned the wheel anew.

Swidden. The first case describes the practice of shifting cultivation, or swidden. The field appears here, then there, then somewhere else, before returning, after a suitable time, to its initial site, where it renews the cycle. Each slashed-and-burned plot creates conditions favorable to the planting of crops. As in nature, the fire releases nutrients in its ash;

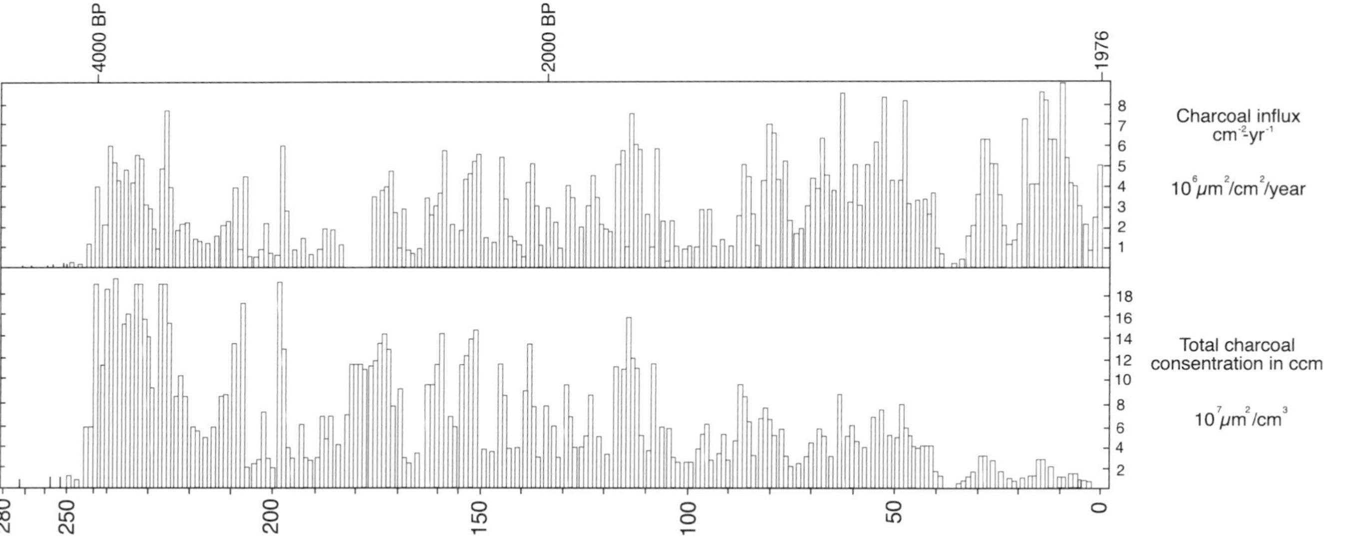

FIGURE 7. Cycles of agricultural fire. Agricultural fire not only relies on cycles of burning but constitutes itself a kind of grand cycle. Consider the case of southern Finland from which some outstanding sediment cores derive, rich in charcoal. The onset of burning commences at the end of the Mesolithic era, overgrown with dense-shade hardwoods like linden (top diagram). The first farmers and herders break it open with classic landnam protocol. The charcoal spikes begin. Thereafter they continue with a pulsing pattern as first-contact landnam becomes domesticated into swiddening. There are variations according to climate, war, introduced crops, migrations, and so on, but a rough rhythm persists. Burning increases towards the end of the 19th century as population and markets both swelled. This was followed by a near collapse in the charcoal record, as migration, industrialization, and state-reserved forests reduced traditional burning. When the carbon record returns, as powerful as ever, it derives from fossil biomass, not forest.

Compare this record with the total charcoal concentration embedded in the sediments (bottom). The largest volume occurs with first-contact fires. The era of intensive, short-fallow burning shows up as a phase of reduced total charcoal because there is less biomass to burn. The era of industrial fire practically removes such burning altogether. Thus the graph frames, for this place, the era of Second Fire. (Source: Bradshaw, Tolonen, and Tolenen 1997, redrawn by the University of Wisconsin Cartographic Lab) purges the soil surface of competitor seeds, microorganisms, and pathogens; opens the site to sun and darkens the soil to help receive warmth; and spurs nitrogen-fixing bacteria. The postfire landscape that in nature often appears bleak and bare is, in fact, readied for its new life. In only a few places could farmers successfully sow and reap by chopping and hoeing alone; the rest need fire.

The dynamics of fire ecology, however, mean that these conditions do not last long. For a season it is possible to plant exotic species and watch them thrive. Then, as with natural fire, the old biota swamps the site. Perhaps, with another, lesser fire, it is possible to extend cultivation for a second season; typically not. By the third year the site is overgrown with indigenous plants and imported weeds. It might be grazed if livestock belong in the system, or it might otherwise be left to flower into fruits, berries, nuts, and other edible or useful plants.

The abandoned plots blossom into biodiverse bouquets. Constant fussing and plucking, however, ensure that the surviving species are largely the ones for which humans have found a purpose. Ethnobotanists have repeatedly documented an extraordinary association of farmed sites with useful species. In central India, some 118 plants out of 121 around certain villages are used; in eastern Amazonia, 106 out of 115. Over many centuries, swidden—aided by hunting, foraging, and perhaps grazing—has weeded out the worthless plants and favored the desirable ones. Equally, as much as 30 percent of the biodiversity of the Pará "jungle" is lodged in the fallows. In such places farming stirs, salts, and simmers the biotic broth.

The first clearing is the most dramatic. While the labor of killing large trees, slashing small ones, or ripping up sod may be huge, so are the returns. First-field harvests are typically many times the norm. In practice, swiddeners try to economize their efforts. Where great forests reign, it is necessary only to kill the trees, not fell them, and then to slash only those trees that produce nothing useful save fuelwood. If the trees are left standing, it is sufficient to girdle or ring-bark them, such that their leafy canopies die and open to the sky and sun. As the land beneath dries, the strewn, parched debris on the surface can burn. Other fruiting trees might be left untouched, or even shielded from the fire. These would survive the burn, provide some shade, and continue to bear nuts, fruits, and flowers. The best surface fuels are smaller-diameter branches and trunks. Some larger boles might be hauled to a sawmill or dragged away for fences or to check erosion. If logs remain in the burn, they will smolder slowly and will have to be rolled across the site, a sooty, laborious task. Then the fire tiller moves on.

The choice of sites is not arbitrary: good plots have ecological properties that make them valuable and that invite a return. They have the right exposure to sun, well-drained soils, and a good stock of vegetation that can be turned easily to fuel. They also follow a social and political logic: there are claims to be made, perhaps fought over. Swidden scouts pick the choicest spots, as a prospector might search streams for "good color." Since plots are small—a few acres at most—the landscape gradually breaks into an intricate quilt of patches worked, shunned, and reworked. The first clearing comes as a shock, the second as cultivation. That second clearing is what makes the system farming rather than plundering. The site regrows; the swiddeners can return, cut again, reburn, and replant. How often they can do this varies according to local conditions, especially the rate at which the dominant fuel rebounds. Virtually any organic material can serve as stocks of fuel: woods, needles, brush, turf, peat, seaweed.

The cycle of burning thus begins the cycle of fallowing, and thereafter follows it. If the time between the returns is too long, the plot loses some of what makes it attractive. Second growth is much easier to work with than old growth. Birch, small pine, and oak are superior to towering fir, linden, and elm. Wait too long between reburns and the labor to prepare the site is burdensome. But if the time between reburns becomes too short, the site's nutrients leach away, the fire burns poorly amid the feeble fuels, and the recycled field declines. Shifting cultivation—true swidden—thus requires large areas, long times, and a politics of land use that allows for footloose farmers.

Fixed-field rotation. Places that cannot meet these conditions tend instead to practice rotational cropping on a fixed plot of land. The site endures, and what cycles is the patterning of plants that grow on it. The sequence begins with the burning of whatever fuels are on the field, or can be brought to it. Then manual weeding by hoe or plow checks weeds; a succession of chosen crops (including if possible nitrogen-fixers) prolongs the soil's fertility; and finally manure from livestock adds nutrients and further stretches the field's agricultural yields. But at some point—perhaps on alternate years, perhaps on a three-year rotation, perhaps longer—the field can no longer support the crops, however clever their manipulated succession.

Yet to say at this point that the field has been "abandoned" to fallow misses a vital point. It makes more sense to say that the fallow concludes the series of crops, that it is being grown to feed the fire as wheat was grown to feed people and oats to feed horses. When the fallow has produced sufficient fuel, it is again burned and the cycle renews itself. In this way the fire ecology of fixed-field rotation matches that of swidden. What differs is that the field and farmer remain, rooted in one place, along with the likelihood for more intensive tillage. Short fallow replaces long fallow.

Quest for Fire, Quest for Fallow

What both schemes share is a yearning for stuff to burn. The search for fallow is endless. It can be mined, hunted and foraged, or grown. First-contact swiddeners often mine it. Secondary ("circulating") swiddeners can forage for it within the landscape's resprouting mosaic. Sedentary farmers, however, have to grow it as they do barley, lettuce, and peas. Only in this way can they ensure a steady supply.

Where fuel exists, fire-fallow farming is possible. When the woods are gone, regrowth and scrub must substitute. If the scrub grows too slowly, coppicing brush can be used, and in fact is sometimes deliberately cultivated between fields or rows so that branches can be lopped and dropped directly on the needy site. Where the fallow is too sparse, farmers must supplement it with branchwood, dung, duff and pine needles, peat, and organic debris of all kinds that they deposit on the field. Burners take care to spread the fuel evenly. They even roll burning logs over the site with long-handled poles. The purpose of the fire is not to dispose of fallow but to burn the plot. Firing is not merely a matter of waste disposal but a means to prepare the site for sowing. A bad burn can mean disaster.

So can an escaped burn. A poorly executed burn may steal into the woods, or into heath or wheat, or into plots prepared by others. Typically, swiddeners surround their fields with a firebreak of cleared mineral soil; sometimes they burn fuelbreaks. The timing of burning helps, too. They can burn when the slash has dried, which is often sooner than (or at least different from) the dormant period of the surrounding woods. In such cases the moisture gradient among the fuels—the slashed plot stacked with parched wood, the forests dank with shade and dew—helps contain the fire. A spot fire that leaps beyond the plot is unlikely to spread or will creep rather than explode. During droughts, of course, this distinction disappears and escapes are many. In organic soils, the depth and breadth of draining determines the depth and extent of burning. Swiddeners can also control the burn by the way they kindle and spread the fire. They may force the flaming front to back against the wind or, if they surround a plot with flame, may compel that ring of fire to draw inward and upward, away from the exposed flanks. And of course burners patrol the flanks. Ideally, field fires are communal events. Elders and custom decide when farmers may burn, and neighbors help neighbors to do the burning. There are plenty of hands to burn quickly, to spread the fire over the whole plot, and to swat out firebrands that might catch beyond the field.

Farming the Fallowed Forest

Swidden comes in endless varieties. Wherever there exists a fuel, there exists the possibility of fire-fallow farming, but among the most interesting variants are those involving woods. Some forests were simply long-fallow swidden: the plots had been abandoned for so long that tall woods had sprung up. ("Jungle," for example, is a Hindi term for "uncultivated land." In subtopical climates, the boisterous fallow led to the dense woods and vines of popular imagery, which made "jungle" a synonym for "rainforest.") Such long-left woods inspired a mix of logging and slashing—hauling off to mill the larger and more valuable timbers, chopping the smaller branches and lesser growth into coarse kindling for another round of burning and planting.

Shorter-cycle fallows, however, were more widespread and their variants many. Some fruiting trees, like mangoes and *mowriïa*, might be spared; others might be deliberately grown with the crops. A good European example involves oak. Farmers cast acorns into the ash along with rye and wheat seed. After a couple of harvests, they would loose livestock, normally cattle, to graze over the site for a few years while the oak continued to shoot upward into poles. After 11 or 12 years, they felled the oak, stripped its bark off to make tannic acid (critical for leather tanning), and minced the debris into field kindling, ready for a new round of burning. As much as 70 percent of Germany's Black Forest was under this regimen past the middle of the 19th century. The system disintegrated under battering by cheap tannic-acid imports from South America.

Pine plantations. In Europe, at least, the economic future pointed toward explicit crops of trees, particularly pine, that could furnish resin and timber and reclaim waste, heath, and rough pasture for productive forests. Fire-fallow agriculture blended into fire-fallow silviculture. Commercial trees, previously sown among other crops, became a crop in themselves. The old practice of burning the site prior to planting endured, but fire entered the scene in other ways as well. Ranks of young conifers, especially, were vulnerable to wildfire. They struggled to thrive amidst greedy grasses and heath, all eager to carry fire. They were surrounded by frequently fired or fire-prone landscapes, full of escaped flames. They had to survive for many decades before harvest, years rich with opportunities for arson, accident, or lightning-kindled fire. Even-aged swaths of pine planted like wheat made an ideal fuelbed to carry wildfire. If the forests were to survive, they would need fire protection.

The most obvious strategy was to actively cultivate the woods, to treat these fields as any other. One solution was to build fuelbreaks into the plantations' design. Another was to intercultivate among the trees, or to create a quilt of coniferous and less flammable deciduous woods; another, to graze selectively once the trees had grown sufficiently that they would no longer be trampled or eaten. Still another method was to gather up small fuels—pine needles, branches—from the ground or lower trunks for use as firewood or bedding. In effect, farmers "weeded" the fallow to reduce the possibility of wildfire. Some places further practiced controlled burning beneath mature trees, a kind of flaming rake to sweep away the piles of hazardous fuels that gathered year after year. Most woods, already subject to special laws and perhaps courts, restricted entry and banned unauthorized fires. They especially targeted swiddeners who valued the trees as fuel, not timber, and herders, notorious for firing woods in order to encourage pasture. And of course the fires themselves were fought.

This often demanded new techniques. A woodland of even-aged conifers burned far hotter than normal heath or wheat stubble or piled cuttings. Far-ranging plantations stoked wildfires well beyond the intensities typical for strip fields or grazed commons. Fighting them required a degree of organization unprecedented for traditional agriculture. Often military troops were dispatched. Without protection against flame, forestry was impossible, yet the decades-long cultivation of the timber cycle was more than traditional agrarian economies could afford. Agriculturalists needed a quicker return. Forestry—and the fire protection that ensured it—thus became increasingly a duty of the state, which could better tolerate a long view. And because woods (or woody plantations) burned so fiercely and resulted in such damage, foresters became Europe's general authorities—the Enlightenment's engineering corps—on free-burning fire.

Forestry and fire. Of all the teeming groups that handled flame or sought protection from wildfire, it fell to forestry to claim free-burning fire as its special charge. In contrast to the others, foresters saw fire more as a threat than a tool. Fire protection meant fire suppression, or, if possible, fire exclusion. Foresters' experience with fire involved long-fallow woods, not short-fallow farms; most fires they experienced were wildfires, not controlled burns. They saw the ceaseless burning by herders and farmers as a menace, not a model. Agronomists had long condemned fallowing as wanton and burning as primitive. Silviculture accepted those values as axioms and sought to farm trees without recourse to fallow or flame.

This was a utopian vision. The belief that they might snuff out all fire remained an ideal, not a practice. Yet like other utopias, this one might be located on islands across the sea, and was. The vision of a fire-free landscape influenced official thinking as forestry reached beyond cultivated plantations into native bush and from Europe to European-colonized continents around the globe. Foresters' practices struck with special force wherever they met true wildlands, places no longer cultivated if they ever had been. Here forestry broke free of agriculture and its fire customs. Here it could attempt to protect the woods for use other than as fuel or fallow. Increasingly the official, the scientific, and the imperial understandings of fire were those which foresters declared them to be.

Flock and Fire: Regimes of Fire-Forage Herding

What holds for farming also holds for herding: a fire-forage regime complements a fire-fallow one. Fire ecology applies as much to one as to the other, including of course the problem of how and why fire fits at all. There are natural steppes and meadows controlled primarily by rain and soil, and there are legions of animals that have munched through evolutionary time without regard to flames. The Earth did not require the fast combustion of flame before animals could eat plants, only the slow combustion of respiration. And there were long times when neither fire nor animals could fast- or slow-burn their way through the plants that piled up. Some of the Earth's greatest faunal irruptions have occurred when huge stocks of biomass were being buried, not burned.

Yet fire has probably hewn closely to forage. Both fire and grazers fed on fine or leafy matter. What promoted one tended to promote the other. The evolutionary emergence of the grasses, in particular, fundamentally changed both fire and grazing and put the two into a curious, often uneasy alliance. While many animals do not need fire to thrive, most have learned to accept fire and have adapted to landscapes for which fire is commonplace. Still, natural fire allowed for plenty of loose linkages and outright gaps. That fire and fauna have become more closely connected in recent times should surprise no one: people have connnected them. Humans domesticated and sought to bond their favored fauna with their favored flames. To sit beside and tend the hearth fire is practically a definition of domestication. Yet that is what the herds and their human herders did, tending the burning hearth of the landscape.

The arrival of fire-wielding hominids changed utterly the ecological balance of power among plant producers and animal consumers. Whatever its origins, controlled grazing relied more and more on controlled burning—for stimulating forage, for expanding and restructuring pasturage, for defending the range's juicy combustibles against wildfire. If fire is removed, herding may shrink to the realm of pets and milch cows. Without those herds the landscape has not only less reason to be burned but less opportunity, because livestock are active agents in shaping the scene. Goats and sheep can split fuels as surely as the farmer's ax and saw, and to the same end. The shepherd's firestick proved as basic as his staff.

How Herding Works as a Fire Cycle

It is possible, through selective burning—by deciding which lands are newly green, which blackened, and which remain still rife with dormant stalks—to control the movement of wildlife; and many hunting societies do this. But domestication is different: the candidates for servant species are few, and it is not obvious that herds rather than individuals are the true source. Likely, animals were tamed by being captured as pups and raised as part of an extended human family. Probably, too, such creatures were few in number. They existed as dogs to serve as sacrifices, then to assist hunting and later herding, as milch cows to supplement the diet with dairy products, as draft animals for transport, as a source of wool or fleece for clothing, and, as necessary, for meat and hide, though this requires that the animal be slaughtered. Typically each family might possess one or a clan several, sufficient for breeding. The animal, after all, must be tended and fed, requiring attention and fodder. Indeed, some societies conduct rituals in which the animal is symbolically adopted into a family, perhaps in ceremonies before the hearth. Certainly domesticated creatures sometimes shared the human fireside.

Such creatures are pets—some only ornamental, some productive. Herds and flocks—a full-fledged pastoralism—require more. The selected species must breed in captivity, must accept the human herder as master, and must be able to forage in ways that agree with human practices. The prevailing theories hold that herding emerged from husbandry, that pastoralism developed after or out of farming, that even where it ranged most widely, even nomadically, the flock continued to orbit around the field. Perhaps so. Certainly the Pleistocene depopulation of megafauna hollowed out a biotic vacuum into which livestock, along with their supporting fires, could rapidly expand.

Fires there were. Unburned forage was often inedible forage. Fire flushed rough pasture and browse; it jolted dormant herbage and browse to life, especially in tropical or subtropical biotas where dried grasses had little protein content; it often forced a reluctant biota to yield forage or to increase what forage it had. Outfitted with the torch (and aided by ax and hoof), people could creature pastures or hold them against a changing climate that would rather grow trees. Without fire—or without the compounding effect of fire and browsing—many humid grasslands would sink beneath woody scrub or ripen into outright woodlands. Tallgrass prairie in North America, Brazilian *cerrado*, South African sourveld—all survive because of regular burning. But generally it is easier to shield old forage than to create new. This is often tricky, since the fine-grained biomass—leafy forbs, blades of grass, tiny twigs—that attract grazers and browsers also suit fire ideally. The surest strategy is to control-burn over land (or around it) before lightning, enemies, or accident do the burning uncontrollably.

Paradigms of Pastoralism: How to Feed Fire and Flocks Both

What complicates the practice of pastoral burning is that the animals are also agents. In the paddock as in the wild, they compete with fire for the available biomass. They can also rearrange biomass into fuel. Goats, sheep, swine, and the lot can open up scrub in ways that favor fire, and are especially adept at trimming woody fallow such that sunny niches for fire remain. Tooth and hoof, in effect, join with ax in gathering fuelwood and splitting kindling. The impact of burning and browsing depends on their timing, which for fire means its seasonality, and for grazing, its intensity. A land that is overgrazed will be underburned. The biomass can exist as either fuel or forage, or both only if its human tenders properly mesh grazing with burning. Likewise, some places that are unbrowsed may be unburnable. Forage, flocks, and fires thus swirl through the landscape in a complex, somewhat atonal dance.

Complicating the scene is the presence of the cultivated field. How closely together farm and flock fit varies enormously. There are farms that exist without livestock, herds that migrate without ties to fixed fields, and fire-forage herding that depends on fire-fallow farming such that abandoned plots regrow into rough pasture and woody browse becomes fodder for swidden. Probably herding developed from husbandry rather than hunting. But herding and hunting often shared similar pastures and certainly relied on similar fire practices. The herder was rarely far from the torch.

Nomadic pastoralism. Begin with the most wide-ranging pastoralism. In such places as northern Africa, Arabia, and central Eurasia, a drying climate brought farming to the brink of failure and shifted the burden of domestication almost wholly to herding. Livestock could exploit lands too arid or hostile to cultivate, and herds swelled in size to provide milk products, wool and hides, transport, and social status. A mixed herding economy of goats, sheep, horses, cattle, and camels could prospect for forage and travel between seasonally or randomly available lands.

Most nomads thus follow the forage, which reflects the fitful rains. Their herds move opportunistically among dune, oasis, and savanna edge. There is scant incentive to burn, though some peoples do. (The Navajos of the American Southwest are reported to have burned individual shrubs to encourage the sprouting relished by their sheep and goats.) But fire here competes sharply with the flocks for biomass, and broadcast burning is rare, when it is even possible. This rhythm breaks when exceptional rainfall years yield a fleeting outburst of growth. Since herds cannot build up fast enough to crop it all off, the drying biomass can then fuel fires. When this happened in central Australia in 1974–75, a huge swath burned, perhaps as much as 15 percent of the continent.

Migratory pastoralism. Where rainfall supports consistent forage, a migratory pastoralism moves through the landscape more predictably. In part, this movement reflects seasonal rhythms, especially when the herds trek up and down mountains or between the coast and the interior. In part, it reflects how much forage is actually present, itself a chronicle of past grazing and burning. Success requires that the practices become regular, that herders act with the same kind of fire-ecology cunning that swiddeners show. They veer into another kind of shifting cultivation, though with animals rather than plants.

In such circumstances, controlled burning can become both common and selective. Burning a few weeks ahead of the herd's spring arrival means fresh forage, and burning behind the retiring flocks (firing the pastoral fallow, as it were) ensures new growth early the next season. Different patches burn as they cure or lie abandoned. The grazed landscape takes on the same fire-cycling patterns as the swiddened landscape. Again, however, the burning is not always essential. There are places that remain grassy and robust in fire's absence, provided they aren't overgrazed. But the same fire power that allowed farmers to leave river valleys has prompted herders to trek beyond natural paddocks and sprawl across vast, fire-forged ranges. Just as farming remakes the fire-fuel cycle into a fire-fallow cycle, so herding converts it into a fire-forage regime.

Typically, the herds move among pastures with the seasons. The pastoral variants are many, though the best documented come from Europe and fall into three broad groups, ranging north to south according to climate.

In the north, the *saeter* system prevailed. Here herds—mostly cattle and goats—traveled from winter barns to summer pastures, and herders, often women and children, tramped with them. The stations became seasonal residences, removed some distance from the core farms. The open range fed the herds and yielded dairy products that could be stored to feed the herders over the winter. The farm, meanwhile, grew fodder to feed the barn-held herd through those same months. Such a scheme is ideal for burning since it separates the paddocks from the cultivated fields. Each burns according to its own cycle, neither overlapping directly with the other.

In the south, along the rocky rim of the Mediterranean, the flocks passed between valley and mountain, a practice known as *transhumance.*

The first furnished winter forage; the second, summer. Pastoralists drove the flocks between those two ranges seasonally, arriving at the one flush with forage when the other was dormant with drought or snow. Transhumance boasted regular corridors of travel, special rules, even state-sanctioned monopolies. Local versions were as many as the landscapes they traversed. In Spain, the herds moved in bold traverses across the *meseta* as well as more tightly between hill and plain. In southern France and Italy, the flocks trekked up and down the slopes between seasonal villages along well-worn paths. In Greece, too often, the herds barely connected with the farm at all; and herders, like satyrs, stood on the margins of society. Similar regimes of one type or another flourished in the mountains of Turkey, Iraq, and Iran. Almost everywhere the route of flocks traced a trail of fires.

Temperate Europe saw mixes of both schemes, along with a surer bonding to the farm. The Alps, for example, experienced a transhumant-like herding as flocks swarmed up the mountain flanks, chasing the spring snows. Regular, *saeter*-like stations existed along the way. Routine burning, sometimes aided by cutting, kept the mountains in pasture rather than woods. Similar practices sprawled over the Carpathians, the Pyrenees, and the Balkans. They both pinned the herd to the farm and unhinged it. The first happened in winter as the flock huddled in the barn, fed by the field-cultivated fodder. The second occurred as summer released both herds and herders from the social and political discipline of the farm and deepened the sense that they were agrarian outcasts, flaked off from the social order. In particular, their fires were said to threaten the uses other groups might make of the landscape. (Immigration could heighten the contrast, as when Basque herders transplanted their practices to the Sierras and Cascades of western North America and aroused public condemnation for wanton burning.)

Elsewhere, when plains replaced mountains, the distance between flock and field shrank. The *saeter* system collapsed into a pattern of cultivated infields and pastured outfields. The infield was tilled intensively, its fallow period defined as one year in three (or less), its crops nurtured with the manure gathered from the tended flocks. The outfield was more varied and feisty. At times it might be swiddened, at other times kept as rough pasture, perhaps allowed to grow into woods. The herds would trek to the outfields daily during the growing seasons, while at night and during the winter they would be housed in barns or pens, where they had to be fed and where they deposited their field-fertilizing manure. The distance between cultivars and livestock shrank, and the ecological divide between field and flock closed. It even seemed to some agronomists that it should be possible to farm the field without relying on fallow or fire altogether.

In fact, the infield could rarely thrive without the outfield, and the outfield sooner or later had to burn. The system shrank or displaced—it did not eliminate—the old practices. The fallow moved to the outfields, so that agricultural fire burned more intensely along the fringes than it did within the arable fields. Instead, the slow combustion of metabolizing livestock linked the two landscapes. Close-herded cattle, sheep, and swine fed on the mast, browse, and coarse grasses of the outfield, all of which tended to rely on regular burning, then returned to fertilize the cultivated infields with their dung and to churn up the soil as draft animals. The apparent abolition of fire was an illusion. Remove the outfield and the infield would starve. Remove fire and the outfield would strangle in its own wild growth.

Herding and husbandry. Yet as populations increased, as better rotations of crops emerged, as agronomic techniques improved, more of the landscape was absorbed into close cultivation. Animal husbandry could replace free-range herding, yields could ratchet up, and reliance on fire and fallow could decline. More of the outfield could be absorbed into infield, perhaps as sown pasture, and one of the three rotated crops could be fodder for livestock. Animal breeding so improved quality that output could rise without adding more land as rough-pastured outfield.

Eventually, a pattern of two distinct fields, one cultivated for crops, one for livestock, emerged. Grown fodder, specific to the animal, replaced rough forage, such that the breeding of livestock had its counterpart in the breeding of special pasture crops. Plants that accepted trampling and invested most of their biomass in their surface growth—traits not typical adaptations to fire—became increasingly prominent, as in the snug bond between dairy cows and white clover. Instead of cycling the flock through the fields, the field was cycled through the flock. Pastoralism ceased to wander, and husbandry replaced herding.

Until recently, however, the rotating fields still relied on rotating fallow, and therefore on fire. More intensive tillage usually meant less fire, but at some point fire had to enter. In fact, most of the Earth did not allow close cultivation on the European model because rainfall was unreliable and highly seasonal, soils were infertile, transhumant mountains loomed somewhere over the rainbow, and fire was endemic. So the fixed plots that were farmed remained small in area. Instead, both field and flock moved; both in some way remained linked; and both looked to fire to propel them along their separate ways and, at the same time, to weld them together.

What They Meant to Each Other

Until recent decades, agriculture has shaped more of the Earth's fire geography than any other practice. It brought to the torch what flame most craved: fuel. There is a sense in which, for agriculture, fire helped swing the ax, pull the plow, and shepherd the flock. In return, farmers and herders struck the spark, stoked the flames, and banked the coals that carried fire to the most forbidding places and kept it aglow. Few sites escaped. Even today farming and herding remain the most common purposes for biomass burning, fallow the most abundant fuel, and the rural scene the most stubbornly steadfast habitat of Second Fire.

But agriculture brought more than raw fire power. It did of course, by accident and arson, allow for more fires to escape and many of these to burn more fiercely. Mostly, however, it disciplined fire into patches and pulses that better suited human society. Increasingly, fire's places were those that people chose for it, and its cycles obeyed the human-selected rhythms of felling, plowing, droving, sowing, reaping. Agriculture further socialized fire, as aboriginal acquisition had humanized it. Agriculture carried fire where it had not routinely existed, changed fire's regimes in places that already burned, and implanted free-burning fire further into the social and cultural order of human existence. It subjected more and more of the Earth's lands to the dominion of anthropogenic fire. There was much more fire than before and it was, by human standards, better behaved.

Few things people do are as complex as agriculture, for its interpretation changes with its context. For fire history, however, the maelstrom of meanings reduces to this, that agriculture fashioned fuels, which could then be burned, which allowed for the sowing, reaping, and grazing that made it possible for people to improve the productivity of old lands and to expand into new ones, and that it created habitats for fire on the scale of continents.

But the deed was trickier than the idea. Agricultural fire proved harder to tend than a controlled fire in hearth or furnace, more thoughtlessly willful, a servant as obedient to blustery winds as to human commands. The untidy field and the sloven paddock practically begged for wildfire. Fuels wobbled between what was necessary to support tilled fire and what invited wildfire. Control demanded a discipline of both spark and fuel. The hand that joined fuel to flame obeyed a mind that ceaselessly puzzled and fretted how best to do so. Not everyone liked the costs or accepted them as inescapable.

The more controlled the fire, the less necessary it often seemed to many intellectuals. Especially where farming intensified, as in temperate Europe, agronomists saw fallow solely as waste and fire strictly as hazard. They reckoned a society's reliance on fire as a measure of its primitiveness, and the scope of free-burning fires as an index of its social disorder. Wildfire most flourished, they reasoned, where landscapes broke down because their human tenders had stumbled when hit with war, unrest, disease, drought, or deluge. For a garden, fire belonged, if it belonged at all, only within the piled debris beyond the plowed furrows.

For such critics—especially common in Europe—fallow was a shamelessly unused field and fire a mere tool, and an unpleasant and unreliable one at that, not an ecological process that humans had tamed and hence had to tend as they might hoe carrots or break an ox to halter. Ideally another, more "rational" technology could, in time, render fire obsolete, as an iron ax might replace a stone one. Meanwhile, the "wastes" could be cultivated, and biomass could pass into compost, fodder, or domestic fuel, not strew a fallowed landscape like so many oily rags. Fire and its hazards—escapes, sooty air, uneven combustion—would vanish into ecological dustbins. Pursued to a logical end, a truly modern agriculture should pass beyond the fire-and-fallow cycle and transcend humanity's messy, awkward, flickering, addictive dependence on flame once and for all.

Such considerations mattered little to pioneering agriculturalists. They could farm and herd only with fire, and saw no reason to withhold the torch in favor of a platonic ideal which, however incorruptible, was hopelessly impractical. Without fire, agriculture would wind down like a neglected clock. The argument mattered only when Europeans became imperialists and could try to stamp their peculiar fire vision onto other lands, and when, with industrialization, the dream of a fire-free agriculture suddenly became flesh and it proved possible to keep fallow buried below the surface and mechanically hide the taunting flames that burned it. Until then, fire and fallow were as much a part of agriculture as seed corn and digging sticks.

RITES OF FIRE

Fire was never far from ceremony. For some rites, fire itself was the focus, but there were many more in which it was simply an enabler. Anything done at night needed fire for light. Anything in the cold needed it for heat. Anyone that required a task performed—an offering burned, incense sent skyward—turned to fire to do it. Eventually, so intimately was it linked to the rituals that fire became no less integral to their symbolism. Long after fire ceased to be worshipped, far beyond the times when it was needed to see, warm, and sacrifice, ritual fire endured as part of the moral ecology of human life.

The origins of fire worship probably date beyond the origins of our species. Fire was too powerful and too mysterious not to be worshipped. Most of the oldest religions have a fire god, and some only a fire god. Even in the Bible the first manifestations of Yahweh are through flame and smoke. Divination by fire, pyromancy, was an ancient rite, and sacrifices were typically burned. The rise of smoke told how the deity received the gift. Certainly the fire god would be immanent at any sacred rite, and certainly the fear of losing him was profound. A lapse of fire for any length of time could be ruinous. The great emblem of this fact, and the fear behind it, was the perpetually kept flame.

The best known are those from the ancient Mediterranean. All are variants of the hearth fire made sacred, an eternal flame that defined family, tribe, and state. It was tended constantly, not allowed to mingle with foreign or profane fires, and renewed only with elaborate ceremony. Consider, for example, the celebrated vestal fire of Rome, which clearly combined religion, tribe, and politics. Vesta was the goddess of the hearth, her shrine the oldest in Rome and the only one that was round rather than rectangular. The fire was overseen by the pontifex maximus, *the chief priest who served as patriarch of the state and representative of the gods. For fuel, it burned oak, the wood most favored by Jupiter. For tenders it relied on four to six virgins contributed by Rome's leading families.*

This practice, too, emerged from the family hearth. As parents aged, it was common to hold one daughter back from marriage to care for them, aptly symbolized by tending the hearth. Just as the vestal fire was the hearth fire elevated to the level of the state, so Vesta's virgins were the homebound daughters committed to its care. Their service extended for a period of thirty years, after which the woman could return to society, her vows discharged. Celibacy was at first a guarantee that the woman would remain in the household, later a symbol of the purity of the fire she tended. Unfaithfulness through either illicit sex or the extinction of the fire brought severe punishment, even to the point of being buried alive. The vestal virgins thus remained under the patria potestas *of the* pontifex maximus*—daughters, not concubines. The* ignis Vestae *was the family hearth fire writ large—purified, perpetual. From it each first day of March citizens renewed their domestic fires.*

There were other ceremonies for which fire was integral. The best known are those of Europe, which collectively make a calendar of fire rites. The oldest was the need fire, kindled during times of distress. All fires in the community would be extinguished, then a new one lit by primitive means, typically by the rubbing of sticks. This new fire would then be carried to all households, and diseased livestock (and people) passed over or between the flames to purge out evil and promote good. Herders added ceremonial fires during the spring and fall (Beltane and Halloween); farmers, fires for the winter and summer solstices (Yule log and midsummer bonfire). Emblems of evil (like witches and warlocks) might be thrown to the flames, particularly during Halloween. Unable to ban the burns, the Catholic Church later absorbed them and added others like the Paschal candle.

But as open flame disappeared from daily life, so did the fire ceremonials, and as the Enlightenment spread, few of the educated elite could see any purpose to the fires at all. They beheld them as blind superstition and witch-burning, not as rites that had evolved out of fire's practical biology, its capacity to purge and promote. Today, when most people in developed nations live in cities, there are few ceremonies of open burning left. What once inspired awe now reeks of the quaint and disreputable. Fire rites have shrunk to votive candles and eternal flames over memorials. For intellectuals, the flame has become sheer symbolism, rooted in an archetypal subconscious. It speaks a deconstructed ecology of culture—words that come from words, rites from rites, symbols from symbols—not as something whose practical effects were known to every hunter, forager, farmer, herder, or anyone else whose contact with flame resided outside books, cities, and TV screens.

Chapter Five

Frontiers of Fire (Part 2)

FIRE COLONIZING BY AGRICULTURE

By its very nature, cultivating is a kind of colonizing. Agriculture converts a biota into a form it would not naturally take and cannot, without constant meddling, hold. But some conversions have gone on for so long and across such vast areas that they have blossomed into full-blown colonizations. They propelled fire into wet scrub, rainforest, swamp, and temperate woodlands, into floodplains and up to mountain krummholz, and did so with such staying power that they sculpted new fire regimes. This was a colonization so mighty that it makes the hominid use of the firestick pale in comparison.

The stories of contact and conversion vary, as they do for aboriginal fire, and for many of the same reasons. Some lands already knew disturbances, had (or had lost) herbivores, had some variety of burning. The greater the contrast between an agricultural landscape and what it replaced, the greater the impact. For lands already marinated in fire and other disturbances, however, agricultural colonization brought only a shift in emphasis, a tinkering of fire regimes. And it left a more subtle record, one often tricky to disentangle into its separate parts. In such places an increase in charcoal alone is not evidence of newly arrived farmers and herders. It may indicate a drought, a disturbance in human society like a war or plague that upsets normal burning, or some longer climatic wave of fuels and storms. The transition from aboriginal fire landscapes to agricultural ones is often one of degree, not of kind. To identify the advent of agriculture requires additional evidence: archaeological, written, the sedimentary residue of pollen from cultigens and weeds, an outrush of eroded soils. In such places fire is not by itself diagnostic. It is too much a part of nature, too integral to the broad-band spectrum of human acts. More fire may not always mean agriculture has arrived. But often it does.

The fire history of agricultural colonization reflects these complexities, yet its venerable history parses, usefully, into three phases. The first lumps together all the agricultural pioneering done prior to the Great Voyages which launched the overseas expansion of Europe. Over millennia, centers of cultigens and livestock emerged. Plants and animals spilled outward in various combinations until by the 15th century agriculturalists had reached all the inhabited continents (save Australia) and were still pushing into new landscapes. Most were variants of fire-fallow farming and fire-forage herding.

The second phase describes the extraordinary expansion of Europe. This slow eruption, originally commercial rather than agricultural, soon became agricultural as trade and empire redirected native farming and herding and as Europeans began themselves, as cultivators, to colonize immense sweeps of land. This process affected all the vegetated continents, and all prior agricultural landscapes. What makes it different from previous agrocolonization is that it linked parts of the world not formerly joined and precipitated exchanges of flora and fauna that could not have occurred under natural conditions. Fire ecology began to span the Earth's biotas, nutrients and species flowing along routes of trade.

The third phase saw agriculture merge with industry, or more precisely, watched as industrial combustion and fossil biomass began to supplement and eventually replace the practices of traditional fire-fallow farming. This phase is yet unfolding. It has burst into select landscapes such as Amazonia and Borneo like a pyric supernova. Yet it has also prompted a process of agricultural decolonization in such places as Europe, North America, and Australia, with profound consequences for fire. In effect, the Earth's fire ecology is reaching into the geologic past for fossil fallow.

How Conversion Leads to Colonization

Agriculturalists burned because they had to. Except for a handful of places, extensive farming and herding were impossible without burning. A controlled disturbance is what made agriculture ecologically possible, creating conditions that did not exist naturally so that imported plants and animals could flourish. Fire was usually a necessary catalyst, especially in lands ever more removed from (and biologically odder than) agriculture's hearths. Nor did it hurt that many of the cultigens and livestock had originated in areas regularly stirred by fire. Farmers burned for the same reasons they irrigated. They needed usable habitats.

Agriculture required fire, fire demanded fuel, and expansion depended on the ability to amass those combustibles. They could be grown, which is what fallowing did, or raw biomass could be converted into combustibles by slashing or girdling, which leads to true fire colonizing. Farmers and herders sought out fresh fuel the way trappers did furs or miners ore.

Agriculture's fire frontier had its own distinctive sagas. Some told of first contact, some piled up layers of storied sediment. The sharpness of that frontier depended on what kind of fire, if any, already resided in a place. The steeper the fire gradient, the stronger fire's effect. The border between aboriginal and agricultural landscapes frequently blurs, and the transition is modest from the semicultivation of a biota by firestick farming to its outright replacement by fire-fallow agriculture, from fire hunting to fire-spurred herding. In such cases, fire undergoes a shift in regime, the size and arrangement of a landscape's patches change, the timing of the burning slides along seasonal scales, and the intensity of the burning both heightens and becomes more predictable. Yet the overall impact of agricultural fire may be subtle. Probably this is what occurred around agriculture's hearthlands, where no clear break separated agricultural burning from the abundant fires that preceded it.

But where the contrast before and after agriculture was sharp, so was the fire frontier. The farther one moved from agriculture's hearths, and the greater the chasm between a landscape's native fire immunity and the vast kindling wrought by agricultural fire, the more such slashing and burning marked the first real onset of hominid fire. The shock could be profound. Certainly the labor involved might be immense, the fires frequently dangerous, and the first-flush yields of crops and fodder extraordinary. The Neolithic revolution could be the biological equivalent of a gold rush. Farmers claimed the prime sites first—agrarian placers, easily slashed and fired—and reaped exceptional outputs. In Finland, old-growth swidden could yield 80, occasionally even 100, bushels of rye per hectare, a recycled swidden only 20 to 30. These were impressive incentives to move on to new lands, even when it meant social isolation and when agrarian colonizers had to wrest those places from aboriginal peoples already living there.

That windfall harvest was not how the story ended. It was enough initially to sow a first crop into the ash and lightly harrow it with spruce branches or to loose small herds into the fallowed browse, but a full conversion required that fire return. The premise was that field and flock would revisit a site over and again. As time passed, the dead, unwanted trees—ring-barked but standing—steadily fell. The desired, fruiting and nut-rich trees survived, the understory thickened with scrub and young growth, easy to chop and crush for the next round of burning. Each cycle of fire became easier, as the landscape converted into slabs and slivers of malleable fuel and fallow. With each cycle it became simpler to prepare the combustibles and control the flames, simpler to seed and harvest. As the cycles returned, the drama of first contact was domesticated. Epic sagas of colonization became the mundane refrains of cultivated crops sprouting one after another.

Stories from the Fire Frontier

Agriculture had its borders. Most were ragged, sloppy, full of spillover fires. The burning adjusted to the winds, seasons, and vegetation of the land. Moreover, unless held by a short leash of fuels, domesticated fires could become feral, and controlled fires rabid. Burning took on the character—the order and chaos—of the roving societies that used it.

What resulted defies easy labels. But the simplest approach is to characterize agricultural colonizing according to how the new fires related to the old ones. Several patterns are of particular interest: where agrarian colonizers brought fire to the land for the first time; where swiddeners and pastoralists overwrote a landscape already rich with aboriginal burning; and where agricultural fire restored fire to a landscape that had lost it. These are, respectively, the fire stories of the far-flung Austronesian islands, of sub-Saharan Africa and the Americas, and of Europe.

A Story of Fire Arriving: Austronesia

When the Han Chinese drove southward and crowded aboriginal peoples and swiddeners to the margins, the ancestral Austronesians left the mainland and took to the sea. They colonized Taiwan, then much of Indonesia, the Pacific Islands, and even Madagascar. During their immense diaspora, they carried with them, or acquired along the way, the pith of a fire-fallow agriculture.

The colonization of Polynesia illustrates what even simple agriculture can do. The great outrigged voyagers carried dogs and fowl (and rats), and in some instances pigs, but for the most part they relied on the slash-and-burn cultivation of taro, the extinction of competing fauna, and the broadcast burning of existing grasslands or freshly fallowed fields. Most of the Pacific islands could burn—with patience or hard work. They featured wet and dry sides, and usually wet and dry periods, some that rode the long rhythms of ENSO. Like embers caught in swirls of wind, Polynesian spot fires blew across the Pacific and kindled island after island.

Each isle, like a ceramic pot, went into a hominid hearth for firing. Some cracked, some were hardened. The shock could be especially dramatic when islands were tiny or lushly stocked with megafauna. On small isles, with less room to buffer the blows, the outcomes were stark—birds suffered, coastal fields burned regularly, and lowland forests churned into long fallow. On Easter Island the stresses eventually broke the biota. The great trees perished, never to return, the soil eroded, and biomass went up needlessly in smoke. Too isolated to receive help, the small island lost the ability to support its endlessly quarrelsome humans, and their numbers too shrank. First contact came probably around A.D. 400, the collapse by 1500. By the time Europeans landed, the land was prostrate, fit for a few sheep, archaeologists, and tourists, in what has become a popular morality tale for the Earth as a planetary island.

The larger islands, having some ecological slack, could better absorb the blows. Yet even the largest, New Zealand (actually two islands), showed the impact. North Island was a typical Pacific isle, volcanic, along the margins of the subtropics, and it underwent a normal Polynesian conversion beginning A.D. 950–1250. Over the centuries the native woods thinned into bracken or thronged into scrubby fallow. The landscape became as dappled with swidden plots as with simmering mud pools and fumaroles. By contrast, South Island was a slab of continental crust whittled off Australia, and it resided out of the tropics. Its soils were leaner, its climate frosty. It proved too frigid for taro and was not remade agriculturally until a cold-weather tuber, the potato, arrived. Instead, the Maori converted large swaths of the eastern, rainshadow landscapes into bracken and particularly tussock grasses by hunting and burning. The Maori's long-voyaged agriculture was sufficient to colonize North Island, but not South, which had to await contact by Europeans who came outfitted with temperate-climate species and long experience in the conversion of cold lands and stubborn soils.

Still, the dry-wind, lee sides of both islands could be burned, and were. Fire-tempered forests could be converted to fern and fallow, tussock grasslands could be expanded, and a score of moas—species of large flightless birds occupying the niches that absent mammals did elsewhere—were exterminated. While buried charcoal reveals that outbursts of burning had occurred long before humans arrived (the biotic equivalent of volcanic eruptions), the real change in regime came with colonization. On the eastern half of South Island, and throughout much of North Island, fire ceased to be a visitation and became a resident.

The most spectacular outlier of Austronesian expansion, however, was Madagascar. The coasting of the Indian Ocean had taken centuries, with lengthy layovers, particularly along Africa. Here the progressively more hybridized Austronesians—call them Malagasy—absorbed practices typical of African swidden and acquired zebu cattle. The large grazers made Malagasy agriculture both different and far more effective at conversion than their Pacific cousins. So did the climate of the island, in which wet and dry seasons blew like a seabreeze. It was a place that aboriginal fire probably could have overrun had it arrived in force. Instead, first contact landed with the heavy ordnance of agriculture.

The earliest Malagasy stormed ashore probably around A.D. 500; another wave arrived perhaps around 1000. The outcome was striking. The Madagascar megafauna melted away before torch and spear as cattle and humans replaced them. Mountain forests broke under the blows of taro-planting swidden (*tavy*). And savannas spread over the central plateau, swollen with flocks of zebu cattle and seas of near-annual fire. A place that, prior to human colonization, had displayed a regimen of fitful, if spectacular, conflagrations steadily shifted into a regimen of regular burning. A landscape that had boiled over from time to time now endlessly simmered. Yet nothing that happened should surprise us. The surprise is that Madagascar waited so long to receive Second Fire. Most places that were prone to burning on this scale people had already reached before they acquired agriculture. In coarse outline, the story is the same as that played out in the Pacific isles, but the shock was greater because Madagascar had a more complex ecology to disturb and the Malagasy more powerful means to upset it.

A Story of Fire Replacing: Old World Africa

Sub-Saharan Africa had no shortage of fire, and had endured anthropogenic burning longer than anywhere else. Yet agricultural fire arrived relatively late. For while much of Africa was ideal for firing, it proved less so for farming and herding on the classic Fertile Crescent and Mediterranean models. The Earth's second largest continent had more fire and less settled agriculture than any other place.

The reasons are many. Africa's native biota, especially its megafauna, persisted after the Pleistocene in far greater numbers than on other lands after human contact. Europe lost its woolly rhinos and mammoths, North America its mastodons and giant ground sloths, New Zealand its moas, but Africa kept its giraffes and Cape buffaloes and much of the rest. Extinction did not, as it did elsewhere, create faunal vacuums for newcomers to seize, nor did Africa propose candidate species for domestication. Likewise sub-Saharan Africa's plants resisted tillage, outside the inland Niger delta and Ethiopian highlands. With minor exceptions, cultigens from the outside world would have to adapt, or the land be changed if possible, to accommodate them.

This was difficult given Africa's old, leached soils; the dense rainforest at its tropical heart; its annual dry seasons and frequent droughts; the sheer ease of firestick farming and hunting; the desiccation of the Sahara from savanna to dune; the absence of navigable rivers and good harbors; the competing carnivores, mostly nocturnal; and perhaps most tellingly, Africa's immensely hostile diseases. It was difficult for cultivars to take root, for livestock to thrive, and not least for humans to enter landscapes infested with fatal microbes. But the place could burn readily, and with fire it was possible to carry agriculture over nature's imposed barriers.

The evidence suggests that livestock began to move southward around 6000 B.P., roughly the time the modern climate (and Sahara) stabilized. Camels appeared in northern Africa; cattle, sheep, and goats percolated down the grassy savannas of eastern Africa. Sheep and goats made it to the Cape of Good Hope by 2000 B.P. Cattle lagged, plagued by the tsetse fly and assorted diseases that passed through seasonal and geographic filters. Later herders brought fresh recruits down the Sudan and into eastern Africa. Domesticated megafauna began to carve spaces from a landscape overrun by native creatures, from wildebeest to antelope to hippos. Fire regimes jostled into a new order.

The breakthrough came when Bantu-speaking tribes from West Africa evolved partial immunity to the worst diseases, notably malaria; devised iron tools, like axes and hoes; and concocted a tropical swidden that also incorporated some cattle, sheep, and goats. They became the agricultural pioneers of the continent, sweeping over Africa much as Slavic peoples did Eurasia. They absorbed most of West Africa south of the Sahel, and then, outflanking the Congo basin, spread boldly south and east. By the 19th century, aside from implacable rainforest and desert, they had reached the Cape. Along the way they had not merely shoved aside

Khoisan peoples, but had dramatically remade the landscape, all of which they touched with fire. The demand for charcoal by which to smelt iron was alone an unprecedented burden on woody savannas.

The great bulk of sub-Saharan Africa, a continent highly prone to fire, now had it almost everywhere: savannas burned for livestock, hunting, and general clearing; forests felled and fired, then left to fallow; marsh, fynbos, karoo, lowveld, miombo woodland, all not only burned but were brought into the cultural order of agricultural prescriptions. Little escaped. Most forest was in fact long-lived fallow. Most grasslands survived by a regimen of grazing and firing. The fynbos of the southwestern Cape, a floral kingdom in its own right with perhaps the richest biodiversity on the planet, a Mediterranean climate for which lightning was rare, burned according to a narrow window of 8 to 25 years.

Another Story of Fire Replacing: New World America

Latin America nurtured probably three great agricultural hearths, and possibly several lesser. The high valleys of the Andes supported one, though one not easily carried elsewhere on the continent. The coastal mountains and lowlands of Brazil sprouted another, centered on tubers and fruits (from cassava to pineapples to papayas). Swiddeners carried this complex to the Antilles, where it hopped across the archipelago. And the churning mountains and lowlands of southern Mexico sustained another, focused on maize, a cultigen well adapted to disturbance and one that could penetrate widely. Maize, allied with beans and squash, propelled a pioneering agriculture that spread to its ecological limits in both North and South America, from New England woodlands to the floodplains of the Amazon.

Its trek to South America is unclear, and the role of swidden (as distinct from the use of raised terraces) uncertain. Maize matured at a time when climatic changes allowed rainforest to advance over wooded prairie throughout the Amazon basin. Perhaps most distinctive was the absence of livestock. Dogs, turkeys, llamas and alpacas in the Andes, guinea pigs—these were the extent of domesticated animals. They supplied meat, some wool, and a tad of transport. Yet draft animals for plow or cart, herds for meat and milk, flocks for fleece, tooth and hoof to assist in the ceaseless fight against scrub, corrals and barns stuffed with manure ready to spread over the fields—these did not exist, and that fact shaped how fuels grew on the land and the purposes and extent of burning. Hunting remained, and the vast and mottled grasslands of the Americas could support broadcast burning on an often huge scale. But the agricultural leverage that livestock could bring as a source of draft power and manure was not in force. More significantly, a faunal vacuum existed that would prove decisive when Europeans dumped their horses, swine, sheep, goats, cattle, asses, and the rest of their animal ark on the New World's shores.

Before then, the old order moved northward. It shuffled overland into the American Southwest (though not into California). Elsewhere it probably spread by boat. It pushed up the Mississippi and its tributaries, sprinkled the South, and inched to the Great Lakes and a little beyond. Likely, swidden combined with foraging and hunting such that some old plots reburned into browse, and bottomland farming was linked with upland foraging and hunting. Swiddens burned intensely; understory forests, if dry, lightly and regularly. Populations fed by maize brought fire to biotas for which natural fire was scarce, if not unthinkable. When Europeans arrived, the fire-fallow cultivation of maize had already broken much of the landscapes of eastern North America. One agrarian regime was ready to replace another.

A Story of Fire Returning: Europe

Putting fire back into a landscape, or over another regime, had become a European hallmark. Agricultural fire was not Europe's first fire, or even its first human-kindled fire. But as Holocene climates stabilized, a seasonality of moisture had blurred into a chronic sogginess that drove fire, and its keepers, to the margins. A vast shade forest smothered flame from the scene. It took agriculture—Neolithic landnam—to pry open the dense woods sufficiently to put fire, and people, back in. Together they pushed against the frontiers of Eurasia.

Thus Europe knew two agricultural frontiers, one that brought farming and herding into it and one that Europe, in turn, sent outward. The first originated in the Fertile Crescent, an ancient hearth for cultivars and livestock. One great wave surged throughout the Mediterranean, leaving as a deposit a complex of cultivation that still exists, before lapping against the mountain borders of Europe's southern rim. Here the ensemble stalled while it tried to cross into the temperate core, as it attempted to adapt from a pattern of winter rains and summer drought to one of summer rains and winter dormancy. Eventually it punched into central Europe, and then with further adaptations, spread to its island

and boreal fringes and eventually thrust in a great wedge back into the vast core of Eurasia.

Over the centuries, colonizing swidden settled into a more rooted swidden, and then matured into field rotations and husbandry. Rough-pastured herds and transhumant flocks became, where possible, more closely bound with the field. Along sodden, arid, or boreal fringes, where farmers could not grow traditional cereals, vines, and fruits, they turned to herding or forestry. In the center, however, fire became a garden tool. It had its place and time: it burned the stubble, the fallow field, the debris trimmed from vine and tree, the outfield pastures. Around the perimeter, fire freely burned with the looser reins of herding, the longer swidden cycles of organic soil and woody fallow, the patches of waste and wild that resisted fixed tenancy. Flame moved through such landscapes because people did too, and when they departed for new lands, it left with them.

That Great Reclamation led to others. What people had once quit, they could later revisit outfitted with new plants and purposes. Such occurred, for example, with the great blotches of organic soil—peat, moor, wet heathlands—that landnam had cleared, then abandoned when the soils became waterlogged or infertile. Those swaths of sodden biomass invited swidden, as much as shade forests of linden and oak had centuries before. By the 17th century the first techniques of "paring and burning" had emerged in Britain. Farmers began by draining the peat with ditches that lowered the water table, then they sliced and stacked the sod for further drying before finally burning it in piles or where it lay on the surface. Into the ash they sowed a sequence of crops until the soil lay exhausted. The land then went into fallow, perhaps subject to some grazing. Eventually the cycle could repeat itself. By the mid-19th century the practice had spread throughout much of temperate and boreal Europe, though not without protest. Agronomists denounced the loss of humus; urbanites, the smokey palls that leaked out of long-smoldering fires. Since it was sensitive to draining and drought, the depth of burning was awkward to control and led to failures of too much or too little fire.

It was pure swidden, nonetheless; and it was characteristic of Europe's ability to recolonize sites with new fire. By the time paring and burning reached its climax, Europeans were busy remaking much of the Earth as they had Europe's core, and were already scrambling into the new regimes of industrial fire. The colonizing fire continued.

Comings and Goings of Agricultural Fire Today

As a force for molding landscapes, agricultural fire holds more than historic interest. It continues today, although equipped with some important distinctions. The first is that Third Fire serves in some way as a catalyst. Modern transport moves goods to markets and people to lands available for conversion; chemical fertilizers and pesticides typically assist open flame. Second, industrialization has also encouraged a countercolonization, a process of agricultural retirement. Tilling fossil biomass can free lands once held for living fallow and can thus reduce the total amount of land under plow and hoof. The story of agricultural fire today as a frontier force is thus one of both advance and retreat. Each may prove equally important for fire.

Recolonizing the Tropical Forest

By the end of the 20th century, the two most glaring examples of advancing frontiers were Brazil and Indonesia. Each nation promoted schemes to move residents from overpopulated to relatively uninhabited areas: for Brazil from an impoverished northeast and modernizing south to Amazonia; for Indonesia from a jammed Java to the outer islands, notably Borneo. Both sought to transform rainforest into farms or plantations, either in commercial or subsistence forms. Both had state sponsorship, not least as a means of linking these fringe regions with the political core. And both have attracted global outrage through their televised burnings.*

Amazonia. Brazil's motives were several, and its means simple. Almost all Brazil's population crowded along the Atlantic coast, often in squalid poverty and subjected, in the northeast, to cripppling droughts. The interior population was sparse; European contact had shattered the native peoples, who never fully recovered, and despite repeated rushes for natural riches (such as rubber), Europeans had never filled that vacuum. Brazil's practical presence was scant, its political grasp feeble in remote regions, and its utilization slight over a domain that it assumed must abound with natural wealth. Frequent calls arose for a regenerating march to the west on the North American model, to have Amazonia

*A good introduction to burning in Brazil and Indonesia is J. G. Goldammer, ed., *Fire in the Tropical Biota* (Berlin: Springer-Verlag, 1990).

serve Brazil as the trans-Mississippi West had the United States. Schemers rallied populist enthusiasm by arguing to bring the people with no land to lands with no people.

Eventually the state intervened. In the late 1950s, Brazil relocated its capital to the *planalto* and cut a road from Brasília to Belém, opening up eastern Amazonia. A military coup in 1964 brought rigor and urgency to the task. New roads were bulldozed through the rainforest, and a mixed crop of settlers and land speculators widened the corridors. Settlers burned to convert the land, then burned to hold it. Smoke smothered larger and larger realms of the subcontinent, saturating almost the entire Amazon basin during the record drought year of 1988. Satellite imagery broadcast the scene—fiery moths eating away at a green carpet—to a global audience.

Fire and smoke made visible a landscape that Brazil had long wanted to advertise, a classic tale of fire colonizing not unlike that which nearly all the developed world had undergone over the past few centuries. But as national geopolitics met international ecopolitics, those images did so in ways that summoned criticism rather than praise.

Kalimantan. A similar story unfolded in Indonesia. The particular promptings were a perceived imbalance among populations, a desire to tap the natural wealth of the larger but less populous isles, alarm over the potential for political separatism, and, vitally, a military-based dictatorship to enforce the state's will. A combination of industrial and subsistence economies swarmed into Kalimantan (Borneo) and Sumatra, especially. Large and persistent fires soon followed—had to follow—as settlers transformed rainforest into palm plantations, swidden farms, and bush pastures. During extreme ENSO events such as 1982–83 and 1997–98 when drought forced even tropical evergreens to shed their leaves, the fires plunged into surrounding forests. A great maelstrom of smoke swirled not only around the Indonesian archipelago but over the Southeast Asian subcontinent, and then onto television screens around the Earth.

Of course fires had long existed on the islands, and people had settled and departed throughout many centuries; of course this process of fire-catalyzed land conversion mimicked those recorded throughout history on every continent and archipelago. The mechanisms were virtually identical to those exhibited, for example, in Russia's Far East and America's Far West. But the differences also mattered—the role of industrialization as an economic prod, the knowledge of historic colonizings and the often compromised landscapes they had left behind as legacies, and especially a global context that made the spectacle visible everywhere.

International attention focused, particularly, on what the burning meant for the Earth's climate. The rapid buildup of greenhouse gases, it was argued, could unhinge the world's climates. This forecast alerted far-removed publics to the fact that a molecule of carbon dioxide released in East Kalimantan merged with those blown from Kenya and Kansas into a global brew. What happened in Borneo could thus influence Britain. Moreover, the culture of the industrial West had fashioned a political philosophy of environmental values that was becoming a green complement to a doctrine of universal human rights. In the past, peoples had agriculturally colonized and answered only to themselves and perhaps those they displaced. Now they had to answer to the world. Agricultural colonizing acquired a visibility and a burden it had not known before.

Not least, that debate over land use often focused on fire—the enabling fire, the apocalyptic fire. Television was the supreme medium for globalizing, and TV demanded action, color, and drama, such that a story of land conversion became primarily a study in abusive fire. Fire imagery became as much a catalyst for political reform as free-burning flame itself for swidden.

Decolonizing Fields

Industry mixed oddly with agriculture. In tropical lands it spurred a wave of colonizing, while in temperate lands it rolled back the old agrarian frontier. Sometimes the upshot meant more fire, sometimes less, the outcome depending on whether the lands were naturally disposed to fire or not. The contemporary Earth offers illustrations of both.

Mediterranean model. The Mediterranean basin—at least its northern arc, in recent decades—is rapidly sloughing off its ancient agrarian skin. Classic village agriculture, with its careful tapestry of arable fields, transhumant flocks, and cultivated orchards is unraveling in the face of market competition and the rush of rural folk to the metropolis in search of richer lives. In places, this inflow of peasants to cities is matched by an outflow of urbanites to seasonal residences in the country. But even where the population more or less stabilizes, the two groups live off the land very differently and thus shape distinctive fire regimes.

This is an ideal landscape for fire. The climate is marvelous, the biota long adapted and disposed to burning, the chronicle of fire as old as its human residents. There is little lightning. Fire prospers because people nurture it. Since this has been ideally a gardened landscape, its fire history closely tracks its human history. People controlled fire by their own fastidious burning, and especially by closely tending the vegetation. Tamed fire kept the feral fire at bay.

But removing or marginalizing the gardeners or replacing them with exurbanites and tourists also removed many of the checks on free-burning fire. Political turmoil has added to the incendiary mix. Between them social restlessness and surging fuels have stoked a rising tide of wildfire. The northern Mediterranean now claims 90 percent of Western Europe's fire load. A new regime is emerging, one that will probably resist complete suppression and will demand, like the region's political insurgents, a say in how the land is governed. Probably this will mean some variety of controlled burning.*

New England, again. New England tells both a shorter and more complete story. Central Massachusetts collapses that whole story into its own. In less than 200 years European settlement had cleared forest from 70 to 80 percent of the land. Then by the mid-19th century farmers and herders began decamping for better lands. A returning conifer forest endured logging as it matured, saw its slash burned fiercely, then suffered abandonment again. Á mixed forest, largely deciduous, regrew in its place. More and more, exurbanites resettled the region.

This was not a place rife with fire history. Without humans—Amerindians, Europeans—fire was rare. The land had to rely on fitful droughts and windstorms such as errant hurricanes to smash the biota into suitable fuels. The Mediterranean could receive fire for a long time after the withdawal of agriculture, but New England could not. As rural practices ebbed, fires became more rare, and those increasingly in the form of wildfires. As agriculture lifted its heavy hand, the landscape rebounded and ripened into more fire-free forms.

Like the Mediterranean, New England got what it wished for, and then wondered that it had not wished more carefully. Removing controlled fire from the Mediterranean only spurred wildfire. Removing

*The literature on the Mediterranean is large; for an introduction, see Stephen Pyne, *Vestal Fire* (Seattle: University of Washington Press, 1997).

agricultural fire from New England, by contrast, left a landscape almost barren of burning. Instead, there sprouted a biota overgrown with woods, often less commercially useful, frequently less attractive (apart from their display of autumnal foliage), and probably favorable to such pests as whitetail deer and ticks. In fire as in geopolitics, a similar logic applied: frontiers had their costs, whether people sought to seize, hold, or shed them.

Chapter Six

Urban Fire

BUILDING HABITATS FOR FIRE

The built landscape is as much a fire environment as forests and fields. It can hardly be otherwise: the hearth, the house, the town—all are designed with fire in mind. Most seek to promote contained fire but, if anything, are more fire-prone than the countryside around them. After all, crowding people together boosts the density of open fires, and cramming structures packs more fuels ever closer. In brief, cities are and have always been fire places.

The same principles of fire behavior and the same pulses and patches that govern other fire regimes affect the built environment as they have wild and agricultural lands. Urban fires behave as terrain, fuels, and winds direct them. Fire cares little whether it burns old-growth slums or ancient spruce, whether it begins from a spilled candle or a lightning bolt, whether firewhirls spawn over ridgetops or around temples. Controlling free-burning fire relies on the same techniques: dousing flames while they are small, dragging fuels away from conflagrations, and setting backfires. So, too, recurring fires trace the contours of fire regimes, which means that an ecology of fire exists for built landscapes as much as for natural or agrarian ones.

The difference between them lies in the degree of control humans have. In principle, our control in cities is absolute. In principle, we can erect dwellings that won't burn or if kindled won't spread or if caused to spread can be contained by architectural firebreaks. In principle, better technology and stiffer social controls could prevent unwanted ignition altogether. In practice, of course, fire has proved inexpungible. Nor, finally, can cities afford to lose it. Even industrialization has only altered, not abolished, burning. Without combustion the city would die.

Hearth and House: Making a Home for Fire

The dwellings that today so shun open burning began, paradoxically, as places to promote it. The earliest shelters—windbreaks, caves, hide huts wrapped around mastodon bones—all held fire, and were in fact often erected with the express purpose of keeping fire alive. The tame fire could not survive day and night, winter and summer, without protection. But neither did people find a building without fire very comforting. So they shielded flame from the wet and cold and kept fuels handy, and as that protected fire flourished, so did the humans who tended it. Fire warmed, dried, and brightened their abode. To be sure, there are tended fires that burn unenclosed by roof and walls, and there exist lodgings without flame. But hearth and house have rarely remained separated for long: together they make a home. Domestication literally began with the creation of a *domus* for fire. The hearth was, as its Latin root reminds us, a *focus* for living.

Where, precisely, did fire reside? It lived variously, as people did. There were special niches like candles and lamps, devices like stoves and furnaces, and assorted appliances that sought to tease out flame's heat and light without the burden of bulk fuel and smoke. But the core habitat was the hearth.

More than dumbly holding flame, the hearth shapes it. Since the fire cannot be allowed to leave its unburnable lair, fuel and air have to be brought to it. How fire's tenders do this, what combustibles they fetch, how they arrange them, how they confine the flames, how they funnel the passage of air—all determine whether the fire glows or flames, whether its heat radiates within a room or passes out a vent, and whether it demands constant fussing or its coals can be banked. The built hearth can influence all of these traits.

Hearths take many forms, and these have evolved. Some changes reflect designs of the fireplace proper, especially chimneys or other ways to vent air. Some testify to changes in materials, the kind of stone, brick, mud, or metal locally available. Even more, others bear the imprint of the fuels available. Wood argues for one design, dung for another, natural gas for still another. And, of course, the purpose of the hearth (or appliance) influences its shape, whether it exists to heat, light, or cook. An open campfire thus requires little care. Details of its design matter only in that the fire not go out or escape beyond its allotted place. But if the fire resides within a dwelling, then the choice of materials, the siting of the hearth, and the character of a vent become critical. As important as fuel, the flow of air into and out of the actual flame governs the fire's behavior. It influences the design not only of the hearth but of the house which encases it.

The hearth has had its time as well as its place. It was where, in the home, the group gathered. A family consisted of those who shared a fireside. Yet cities have always treated it warily, and industrialization has sought to abolish it altogether. More refined fuels and better materials caged it into Franklin stoves, then lodged it into furnaces before exchanging it for electricity and gas. Once the very symbol of the family, the hearth has become merely decorative. In modern life the hearth remains in the heart rather than the house.

For all their commonalities, the built environment has something that wildlands don't: the room. Structures burn room by room (or from the roof down). The behavior of fire within a single "compartment" is the fundamental unit of urban fire analysis. In a room, a fire undergoes a life cycle of rapid growth, mature development ("full involvement"), and decay ("smoldering"), and may not spread beyond that single unit. By contrast, in a wildland setting, this "cycle" of combustion describes only the flaming front as it passes over a given spot, with most of its heat lost in the winds. By being more or less confined, however, a room can trap gases and smoke; heat can amass quickly, even explosively. Equally, a room fire can burn out as rapidly as it builds up.

The core difference between burning indoors and out is the presence and flow of air. Oxygen saturates wildlands and fields, and only the most extreme firestorm can—temporarily—empty a site of oxygen. The fire creates gases and forces air to flow, and winds pour into and around the fire. Almost never does lack of oxygen limit burning. But oxygen is, for a room fire, critical. Like a candle placed under a box, a fire in a closed room will soon devour its oxygen. For a while the room will be filled with searing, ready-to-combust gases as heat continues to dissolve coarse fuels into vapors. If no fresh air enters, the fire will gradually die and the gases cool. But if enough oxygen is on hand and if the heat radiated from that trapped cloud of particles, soot, and gases exceeds the ignition temperature of all the exposed surfaces in the room, everything may burst suddenly into flame—a flashover. Or should oxygen pour into a room bloated with trapped gases and soot—say, by a door or window thrown open—then the mix may explode in the wild rush of a backdraft.

Thus a room fire behaves more like a hearth fire than a forest fire;

the principal reason is air flow. While wildland flames move with the wind, the behavior of a room fire follows the ventilation flow within a building. By widening existing vents or creating new ones (for example, punching a hole in a roof, akin to opening the damper of a chimney), it is possible to control a fire's behavior—to regulate its rate of combustion, to shunt its path of spread, to dampen the potential for backdraft. Confinement also means that water can be more effective in rooms than in woodlands. Water shot into a compartment absorbs heat and becomes steam, which then spreads exactly as combustion gases do, and thus helps smother them. All this is fundamental to fire control in structures. None of it is possible in wildland or agricultural settings.

The logic of fire protection in the built environment is thus to confine a fire to a single or handful of rooms. The ability to do this by architectural design and proper choice of materials is one reason why it is possible to erect large structures. Firewalls and firedoors, room complexes that pass air in particular ways, fire mains and extinguishers where they can be used while a fire is lodged in a solitary room—all have allowed for higher density buildings. Like bulkheads and pumps on a ship, such devices prevent a single-room fire from spreading. New codes that mandate automatic sprinklers carry this logic further, such that fire protection may become essentially automated. In modern cities, structural fires rarely escape byond a solitary building, save on occasions of social chaos such as riots and wars. Urban fire departments define a large fire as one that involves many people, typically in high-rises, not a fire that romps over large areas.

The hearth and the room together make a house. How they fit one to another depends on what materials exist for building, the purpose of the shelter, aesthetics, and the kind of fuels handy. Is the dwelling temporary or permanent? Is the primary building material wood or stone? Is the climate cold or hot, wet or dry? The fireplace (or fireplaces, around a common chimney) might stand in the middle of the dwelling or along one wall. Houses made of wood and thatch must vent their chimneys carefully, lest stray sparks ignite the roof. Places with long summers often seek to separate fire at least seasonally, perhaps with a detached kitchen, while places subject to long winters try to capture heat as fully as possible throughout as much of the dwelling as they can. A yurt burning dry dung has different concerns than a log cabin burning split wood.

While the hearth must not itself burn, the buildings encasing it like a vast windbreak can. Made of combustible materials, stocked inside with cloth, wood, paper—the more they resemble wildlands, the more they burn like a wildland fire. Daub-and-wattle huts combust like forest windfall; thatched-roof cottages, like sedge patches and prairie; log cabins, like timber slash. The same principles of heat transfer and fire behavior apply to dwellings as to woodlands. Radiation and convection are more important than conduction; fine fuels combust more rapidly than coarse ones; and heat and smoke rise, so fire spreads faster up than down. Thus walls and ceilings burn more readily than floors, and flames race upstairs rather than down. Spotting casts embers far from flames, so fires leap from thatched roof to thatched roof. A fire is a fire, whether in a hearth, a house, or a prairie. A common chemistry works through all.

Built to Burn: A Fire Ecology for the City Combustible

Yet fires do not always remain in single rooms or on solitary roofs. They spread among buildings, and in the past they have done so with a regularity that suggests it makes sense to speak of urban fire regimes. Compartment fires became conflagrations, no longer confined, racing through structures as through dense woods. They are rare, as crown fires are in wildlands; but there is every reason to consider wholesale combustion as a fundamental disturbance that has shaped cityscapes according to the same logic that applies to other landscapes. So, too, fire affects cityscapes by its removal. It is no accident that urban renewal projects began as conflagrations ended.

After all, until recent times, cities were fundamentally rebuilt wildlands—composed of similar materials, drying and wetting with the same rhythms of drought and deluge, obeying cycles of youthful exuberance and overgrown decadence. Urban fire burned at the same times, according to the same principles, and with patterns akin to those of wildland fires. Until cities were built with stone and brick instead of wood, until slate and tile replaced shake shingles and thatch for roofing, until streets widened instead of narrowed and were paved with cobbles instead of planks, cities blazed like the forests and prairies from which they were made. The Russian village burned with nearly the same regularity as the slashed swiddens around it.

As cities evolved, so did their construction materials, and so has the spectrum of urban fire behavior. Three environments now characterize urbanized landscapes: an urban core, a suburban fringe, and an exurban periphery. The urban core is the classic city, its density the outcome of its enclosing walls. The suburban fringe was originally that urban-like frill that existed outside and below the city walls. More recently it has come to dominate the metropolitan scene, sprawling wildly outward by the pressure of rapid transport and especially the automobile, more a frontier than a fringe. The exurban periphery describes the still-more-remote scattering of urban fragments, even communities, across former agricultural lands and into wildlands, an urban recolonization of a once-rural landscape. Each of these urban environments has a characteristic fire behavior—each its regimes, each its evolution.

Urban Core: Nuclear Fire

The core boasts the classic urban fires of history: the conflagrations of ancient Rome, the London fire of 1666, the 1812 burning of Moscow, the 1923 Tokyo conflagration, the relentless combustion of cities from time immemorial. Such fires behaved much as wildland conflagrations, and for identical reasons. What made them dangerous was that they always had heaps of fuel and plenty of scattered flame. The trick was to keep the two apart. Codes for buildings were ignored, but those for human behavior proved surprisingly strong. Citizens had to show a self-discipline with regard to fire that they showed for little else. Arson, not surprisingly, was a capital offense.

Yet fires would inevitably break out, and big fires tracked the presence of piled fuel, favorable weather, and flame that had slipped its social leash. Typically, conflagrations swept the older, overgrown, slummier sections of the city, close-packed with buildings and stuffed with combustibles, a congestion of fuels that could burn with exceptional intensity. Of special note was roofing: covering dwellings with grass, thatch, planks, or sod meant that roofs burned like prairies, woods, and organic soils. The kindled roofs accepted sparks easily; fires bounded from rooftop to rooftop.

Still, even city quarters heaped with dwellings like slash would not burn if they lay under snow. Big fires thus required that those fuels be in a condition for burning, parched by seasonal dryness or drought, and that powerful winds aid the flames, pushing fire through buildings, billowing flames over and across streets, and lofting sparks well beyond fire brigades. The great London fire of 1666 burned under the impress of a dry east wind that reversed the normal westerly flow of Atlantic

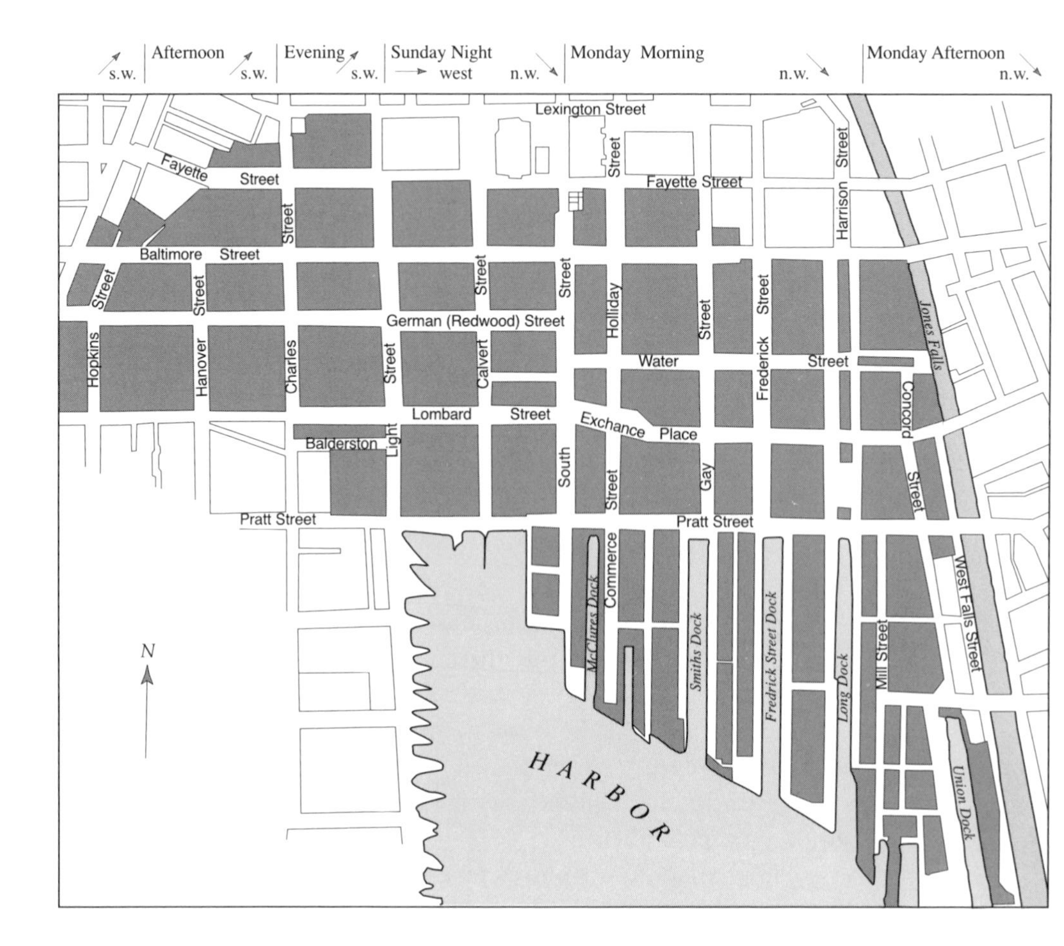

FIGURE 8. Fire in the big city. The last major, purely urban conflagrations in the United States were those at San Francisco (1906) and Baltimore (1904). The San Francisco fires showed complex fire behavior because, over several days, they burned amid local winds and according to local terrain and of course housing (which provided fuel). The Baltimore breakout occurred under the impress of a single dominant factor, the passage of a cold front.

The map tracks both the wind and the burned city blocks through the afternoon. In advance of the front, winds freshened from the southwest and drove the flames northeastward. With passage, the winds began veering from the west and finally from the northwest. The flaming front followed those shifts in lock-step. The presence of the bay sharpens the lesson because it prevented much flanking spread as the winds made their great gyre. A 50-foot wide canal finally broke the fire's progress. (Source: Lyons 1976, redrawn by the University of Wisconsin Cartographic Lab) moisture. The fire burned out a rough ellipse, framed on the south by the Thames. The Chicago fire of 1871 was part of a vast complex of fires throughout the region and ripped through the city with the shifting, blustery winds of a dry cold front. The Baltimore conflagration of 1904 exemplifies precisely the approach and passage of such a front, the flames driven first north, then southeast. The Hamburg firestorm of 1943 arose amid an unstable atmosphere, with gusty winds along the surface underlying calmer winds aloft, exactly the profile typical of blowup wildland fires. Even terrain mattered. The 1906 San Francisco fire rushed up slopes faster than down, gobbling packed, wood-frame townhouses on the hills in minutes. The great fires only ceased when the wind dropped or the fire met a barrier too broad for windblown sparks to vault.

Given their extraordinary hazards, it is remarkable that any cities were left standing. There was no way to exclude fire. Without combustion, the city would cease to exist. Nor until recent centuries was there much success in constructing cities out of fire-resistant stone, slate, and brick in locations where wood and thatch were abundant. Fire-inspired building regulations date back at least to the Code of Hammurabi and have been ignored ever since, especially in the older slums, particularly ripe for burning. No controlling authority with sufficient will existed to enforce edicts over long periods. That changed when fire insurance brought building codes under the regimen of capitalism and the discipline of a market, and when industrialization succeeded in dispersing cities and in mass-producing building materials that were relatively fire-resistant. As fire protection improved, so did insurance, and the two together, embedded in codes and outfitted with fire-retarding bricks, steel, and stucco, have remade the environment of urban fire.

Instead, fire strategy focused on prevention, on ensuring that the legion of open fires had someone to tend them, that pots of water were handy in case of accident, and that some rapid response was possible. For example, fires were to be extinguished or covered at night, the origin of the *curfew* (from the French, *couvre-feu*). Special wardens patrolled the streets, especially in the evening, to ensure compliance. Alarm bells would rouse citizens to aid. Even in Augustan Rome special corps of firefighters would rush to the scene of an outbreak. Until cities were made differently, however, their capacity to stop large fires was slight.

Fire control remained primitive. If the fire was tiny, it could be attacked directly with water or blankets. But if it escaped more than a few rooms, if wind could spread the flames outward, then it had to be fought on the same principles as a wildland fire. Firefighters created fuelbreaks: they emptied houses of furniture and fuels; they demolished adjacent structures; they stripped roofs of their combustibles, typically with the aid of hooks and ladders. In extreme cases, they set backfires. Only in the last two centuries have internal combustion engines made effective pumps, and have water mains latticed cities with plumbed reservoirs that can be tapped for emergency use.

But in general it was social restraint that checked fire. Indeed, it is astonishing that cities were not constantly aflame when they were warmed by open fires, lighted by candles, powered by hot forges and kilns, littered with trash fires, paraded through with dripping torches, and occupied by the careless, the ignorant, the young, the malcontent, the deviant, all of whom had free access to fire. Such control, however, broke down during riots and revolutions. Arson stands often as the very image of malicious unrest. And internal control collapsed during wars as besieging armies deliberately exploited a city's vulnerability to fire. Here roofs were, once again, the point of vulnerability, and a succession of ingenious devices sought to fling incendiaries onto them. In response, the city might peel back roofing from the zone of assault. If fires started, however, either during the siege or after, the fire-defenseless city would roast over its lavish fuels. The material history of cities is often a history of their fires, which were largely a record of social conflict.

Warfare—from either foreign invasion or internal insurrection—dominates the recent history of urban conflagrations. Block-buster bombs and incendiaries have brought slashing and burning to cities on a huge scale. With aircraft and missiles, free-burning wildfires have returned to even the modern metropolis. During World War II, mass fires gutted Hamburg, Dresden, and Tokyo. Nuclear weapons are, as Hiroshima and Nagasaki demonstrated, fire weapons of enormous power. Exploded at high altitudes, they can send out shock waves that shatter structures and infrared pulses that can irradiate those smashed fuels into flame.

Surburban Frontier: Fire's Middle Landscape

The walls that defined cities also confined them. That only changed with the advent of industrial combustion for transport. Steam locomotives and especially gasoline-powered autos have allowed the suburban fringe to dominate the recent geography of urban landscapes. This dispersed settlement has had mixed consequences for fire.

In places relatively immune to flame, it has reduced wildfire. There are lower fuel loads, more firebreaks, fewer open flames in shops and forges, greater attention to fire services. As the 19th century developed and steam allowed suburbs to push outward, a "fire gap" widened in Europe and America. Previously, burned area had been proportional to the size of the city. As cities expanded, so did their fires. Increasingly, however, less burned. The reason was industrialization. Its pyrotechnologies substituted for traditional fire practices, new materials for construction became available, and rapid transport allowed for cities that spread outward rather than back in on themselves. While controlled combustion remained as vital as ever, it was embedded in machines or dispersed to power plants on the outskirts or absorbed into electrical or gas appliances that eliminated the need for open flame. Not least, it made possible more effective firefighting machinery.*

But suburban growth has also encouraged fire where houses nestle in places disposed to fire. Where the city pushes dwellings against and into landscapes lush with vegetation, where the urban frontier spreads rapidly and disturbs widely, where it outstrips fire protection, where it scrambles natural and urban fire regimes, where climatic conditions favor burning, then sprawl can become a formula for wildfire. In the industrial world, what passes for "urban" conflagrations center on just such scenes—from the urban bush of Sydney to the tourist-cluttered slopes of the Côte d'Azur to subdivisions crowded amid the chaparral of Sierran foothills. A profuse "natural" growth connects what suburbanization otherwise tends to split apart. More incredibly, many such communities sport—even promote and advertise—wooden roofing. Flames that can't spread over lawns and through orchards simply jump from housetop to housetop.

Exurban Fringe: Fire's Outer Limits

Still farther beyond the city's outskirts, the proportion of built environment shrinks, and the proportion of wild or rural landscape expands. The dominant fuels are natural; so is the dominant fire behavior. Small clusters of wooden houses are no different from windfall or slash and are simply swallowed up in the larger rush of a flaming front. Here, the direct influence of the metropolis is slight. The purpose of many such

*See L. E. Frost and E. L. Jones, "The Fire Gap and the Greater Durability of Nineteenth-Century Cities," *Planning Perspectives* 4 (1989): 333–347.

communities is, in fact, precisely to reside within a place that appears natural.

The indirect influence of the metropolitan city-states, however, is immense. More and more, urban values, urban politics, and urban fire practices are restructuring their hinterland and backcountry at ever more far-flung sites. They are redefining land use, often to more recreational purposes. And they are establishing a de facto fire protectorate over vast outlands, for free-burning fire seems to have no place in these imagined worlds. The most obvious medium for mingling the urban and the exurban is through their shared airshed: smoke in the countryside competes with smog in the city. Eliminating wildland and rural burning is, for urbanites, a means to shore up their own degraded air quality. And they have the clout to force those choices: cities are where social power resides. Control over what fires can burn, and where, is shifting from the rural edges, where flames exist, to the urban center, which views them askance from afar.

How to Build a Fire Regime

Cities as nested landscapes. There is a hierarchy to urban fire, more definite than in wildlands. The basic unit is the room, buildings consist of many rooms, and cities are made of many buildings. Each has its peculiar dynamics, each has links that join it to the others.

The room has its own enclosed ecology. If it is dry and warm and stuffed with combustibles, a fire can burn within it regardless of what the weather and fuels are like outside. A room can burn at times when rain or snow would quench any outdoor flame. There is little reason, in fact, that dwelling fires should correlate with seasons apart from people's countercyclic reliance on domestic fire for heating during the winter or through the rainy season.

But once a fire leaves the room it falls under another order, the same rules that govern wildland fires. An urban fire will not spread readily if a wildland fire, under similar circumstances, could not spread. A building may burn to the ground, and take those adjacent to it, but the flames will not push further. If, however, urban fires behave more or less as wildland fires do, then they should also burn with similar pulses and patches. They should organize themselves into fire regimes, and they should exhibit a fire ecology. In fact, most do.

For some cities, located in fire-prone landscapes, this fire-likeness between city and country extends to common fuels, a collective climate, and even a shared source of spark in lightning. There is little difference between the timing and frequency of such urban fires and those of agricultural or wildland burning. Most urban fires, however, have causes more likely coupled to business cycles, political elections, social traumas, and random events like revolutions and wars than to seasonal rhythms, hunting, or herding.

History and geography, nature and culture, regularity and accident—urban fire results from their often eccentric collisions. Economic depression and boom may be more important than drought and deluge; a race riot more critical than dry lightning; widespread corruption of building codes more decisive than high winds. Natural events may interact with urban landscapes in ways unthinkable in wildlands. San Francisco and Tokyo burned after earthquakes scattered sparks and shattered water mains. Dresden disintegrated after Allied aircraft dropped 650,000 incendiaries that kindled a mass fire the February countryside of Germany could hardly have sustained.

Common causes: city and country. Still, some common considerations apply to both city and country. Scale, for example, matters. The ecology of cataclysms applies equally to city and wildland in that big fires do more work than small ones. Timing matters, too. Cities burn when ambient conditions favor fire of any kind. Typical villages burn more during the day than at night; most often during that part of the day when temperature is highest and relative humidity lowest; most frequently during the same seasons that allow free-burning fire in the landscape around them. Commonly, the dominant season occurs when dryness is greatest and strong winds are most likely. Villages tend to burn under the identical conditions as their surrounding fields and thus at the same time.

Another commonality is fuel. Village and field may burn according to similar regimes if they consist of the same twigs, grasses, thatch, and logs, only differently assembled. That arrangement is critical. Are buildings clustered or scattered? Are there many vacancies between them? Do woods and shrubs fill those gaps? Is access easy or difficult? Oddly perhaps, or not so oddly, the same principles that govern the fire-protection system of a pine plantation apply equally to cities. (They were as often ignored for the one as for the other.) In both city and country, too, the very new and the very old fuels are the most common locales for large fires—the new because the site is disturbed, the town freshly hacked out of wilds or erected rapidly, its social order inchoate; the old because slums compound and crowd fuels and thus suffer a neglect in social care.

The ecological upshot is that fire recycles. Historically, villages simply rebuilt themselves in the same fashion they knew before the fire. In urban centers new districts rose out of the ashes of the old in a kind of fire-crafted mosaic. Linnaeus observed that he could reconstruct the history of the Swedish towns he visited by their fires, each of which led to a grouped rebuilding, much as one could date a forest through its stand-replacing burns. There is, in brief, a kind of structural succession that mimics that of nature. In classic cities as in wildlands, burning proves stubbornly conservative: it restores—renewed—the previous scene. So long as cities continued to rebuild themselves on the same pattern and with similar materials, they experienced the same kinds of fire. Those recurring fires made a fire regime.

Social ecology of urban fire. Yet cities changed over time, and their fire regimes with them. The same forces that broke down their confining walls also carted in new materials like brick and new designs that discouraged conflagrations and that built into urban geography the capacity to halt those fires that did start.

The most hazardous times were, again, the old and the new—those periods when fuels had heaped to unhealthy levels, those eras when a society was awash with change, when it unpredictably mixed flame with fuel, when it proved incompetent to impose a social discipline. Fires struck hurriedly erected new towns more often than established ones, transient more than permanent towns. In the American West, it was routine for agricultural settlements to burn during their pioneering stage, when the surrounding lands were ripped open and fuel lay ripening in the sun. For logging, railroad, and mining villages, however, it was common to burn and reburn, perhaps half a dozen times before they were either abandoned or fixed. The chronically unsettled landscape around them, made more prone to fire, was a contributing cause, but so was indifference to creating a stable society that could enforce codes, and prevent and fight fires as they broke out.

What this underscores is that in the fire ecology of cities there is no way around people. The city exists only because of us, and for our use. Most wildlands know anthropogenic fire, but many would still have fire if people left. That is not true for cities (even admitting the number of cathedrals especially that are struck by lightning). In the city, almost everything that happens with fire happens because of people. People start fires, people stop them, people stack fuels, people haul them off. People decide what fire practices they will accept, and what fire regimes. Fire ecology is human ecology. If the keeper of the urban flame goes berserk, so does fire. What in nature happens from extraordinary winds, freak dry-lightning storms, or dramatic droughts happens to cities from wars and riots. They are as much a part of urban fire ecology as hurricanes and earthquakes. In their absence, officials have to substitute controlled destruction in the guise of urban renewal. Fire burns in the cranes that swing wrecking balls and the dynamite that levels skyscrapers.

The Eternal Flame Invisible: Fire in the Industrial City

However large they loomed as administrative, economic, and cultural sites, cities were tiny in area. Even Rome was a speck on the Italian landscape. London—the largest European city for centuries—hardly surpassed more than a few manors. As a fire habitat, the city was paltry and its ability to influence fire practices outside its walled shadows was limited. All of that changed with industrialization.

The town grew into a metropolis that stretched to the horizon. By A.D. 2000, metropolitan Los Angeles was larger than Crete, Buenos Aires broader than Yosemite National Park. Nearly half the planet's human population lived in cities. Today, the hydra-like city-states have become hot spots in the global tectonics of combustion. Satellite imagery of evening lights shows the contours of cities, the lines and fields of fire of an industrial Earth. What especially matters is that these emerging city-states control the throttles of social institutions, not only of politics and economics but of environmental philosophy and national identity. Questions about what kind of fires should exist are increasingly decided in urban centers based on urban values. The modern city's fire reach extends far beyond the range of its municipal fire department.

The larger meaning of fire's ecology in cities is unclear. Urbanization had created one fire ecology, industrialization another. Neither is well understood or sharply bounded, yet they have now become joined. Modern cities remain fire-driven ecosystems. Fire's influence is everywhere, yet fire is almost everywhere invisible. Its fuels flow as liquids and gases; its combustion occurs in special chambers and machines; its power is transmitted, often over vast distances, through electrical wires. The fire-resistant building materials—brick, cement, and steel—that dominate modern construction have already passed through the flames, though these be the forges and kilns of industrial pyrotechnology. Cars, trucks, buses, motorcycles, tractors, backhoes, bulldozers, graders, generators, lawnmowers—the urban landscape overflows with a mechanical fauna that feeds on fossil fuels. Traditional lines of fire trace streets, and equivalent fields of fire lodge in factories and parking lots. Shut down combustion and you shut down the city. Like a strange attractor of chaos theory, fire's threat haunts every room and corridor, every multistoried building and mall, the plumbing of water mains and sprinkler systems, the wiring of alarms, the design of building exits, the siting of emergency services. But open flame itself has virtually vanished. Like a black hole in space, fire has shaped everything around it without itself being visible.

It had to be so. Without more robust control over flame, the modern city could not exist. The thickening metropolis would burn as fast as it was built. The solution lay in engineering, design, and low-flammable materials, but behind them all stood industrial combustion. The ecology of urban fire, once squeezed between stone walls, now splashes outward. Industrialization has hustled fire from the city's center and pushed it to the fringes. An exception is the oft-times dead urban core, where abandoned lots, empty buildings, and crowded tenements invite fire and arson, but their hazard comes precisely because they are not lived in and codes are not honored.

The general trend is that, as cities have grown outward rather than inward, they have unpacked fuels, found flame-free applicances, and squashed open burning. The fuel loading of the suburban environment thins relative to the urban core. There are wide streets, shade trees, and watered lawns. Society is stable, codes enforced, and fire services active. Still further out, however, the exurban scene sheds these features. If fire was common before, it remains common, and may in fact sharpen into more violent forms. Instead of trimming and dampening vegetation as typical suburbs do, the exurban scene allows it to overgrow and then sticks combustible houses in its midst. Low-intensity fires that might, in the past, have crept and cleansed the surrounding lands disappear. Conflagrations, from time to time, take their place. In these circumstances, the urban "fire gap" that had widened with the sprawl of suburbs collapses; wildlands and cities are scrambled into an ecological omelette. Just such an "intermix" fire scene has become, for most industrial societies, the dominant wildfire problem.

The fire benefits of the industrial city are many, when (and if) the city can make the full transition. The extinction of domestic fire, in particular, has advanced public health. Poorly ventilated stoves, smoldering dung and debris in house and yard, inversion-capped crucibles of morning smoke all invite serious air pollution and all can be scotched by industrialization. Of course, industrial fire blasts out plenty of contaminants of its own, and cities with the worst air quality (like Mexico City and New Delhi) achieve that distinction by compounding the worst of both regimes, by mingling the nitrous oxides, aerosols, and fly ash from burning fossil fuels with smoldering cooking fires of wood or dung and with agricultural burning, all of this pall hovering close to the places people live. The assumption is that such a state is transitional. That may prove true.

What is gone from the industrial city is fire itself. What has been lost is the daily interplay between people and flame. Instead, industrialization has replaced biofuels with fossil fuels. For the open-flame forge it has substituted new technologies, and for hearth and furnace new combustion chambers. It has even sought to supplant fire's ecological processes with programs of urban redesign and renewal. It has hidden fire's ecology in machines, so that people know ignition through the keys that crank their automotive starters; know fuel through the gasoline they pump from one tank to another with barely a spill; and see fire's effects indirectly through grungy air that hovers around skyscrapers. They regard open flame as at best ornamental, suitable for ceremonial display, but otherwise a nuisance and always a threat. "Learn not to burn" is among the earliest warnings proclaimed to urban children. Remove non-industrial fire and the city would continue, but remove industrial fire and the city would stop. That observation speaks volumes about the relative power of the two fire ecologies.

Before the end of the 20th century, industrial fire was outburning its combustion rivals. Yet by merging with the city, it further leveraged its power to shape the planet. With the world's population sucked more and more into the gravitational pull of industrial city-states, the fire regimes most people (and most officials and intellectuals) know are those they experience firsthand in urban clusters. Here—not in fire-flushed hunting grounds, on flame-sodden swiddens, or over glowing hearths—is where they learn about fire. And here, in shaping fire knowledge, is where urban centers hold sway over the larger realms of fire. Those great city-states of an industrialized world now dominate political, economic, and cultural institutions. Much as industrial combustion is replacing biomass burning, so urban fire standards, practices, and institutions are replacing those that prevailed earlier. The fire regimes of the cities have become, improbably, a norm for all landscapes, allowing the metropolis to change the fire regimes of its hinterlands, proclaim fire protectorates over remote outlands, and mold how urban citizens understand the place of fire on Earth.

Yet cityscapes remain a special habitat and their fire regimes both recent and anomalous. Cities can substitute one pyrotechnology for another to the extent that their ecology, like their structures, is built. One can replace a wood-burning stove with a gas appliance; thatch with tile or tar shingles; burning trash with gasoline-powered sanitary trucks and a landfill. It works. But nature cannot be crafted so thoroughly, and large natural estates cannot tolerate such tinkering, for fire burns in them not as an engineering tool but as an ecological process. There is no substitute for flame rushing over a prairie as there is for the flame under a cooking pot. The contained combustion of chain saws, woodchippers, and front-end loaders does not equate ecologically with a crown fire. The attempt to ban free-burning fire in extraurban landscapes is one of the most significant outcomes of urban industrialization. The once-walled city needs a new boundary, a biological border, firewalls to separate an unquenchable Third Fire from the others.

Chapter Seven

Pyrotechnics

FIRE AND TECHNOLOGY

In nearly all myths, when people get fire, they move beyond the rest of creation; they become distinctively human. Aeschylus had Prometheus proclaim that by bestowing fire on humanity he had invented "all the arts of man." That's a claim as reckless as it is bold. But it is certainly the case that humans are tool users, that fire is among the oldest of human technologies, probably the most pervasive, and likely the most enduring. Since they first met, people and fire have rarely parted. Together they have crossed deserts and glaciers, passed into rainforests and oak groves, sailed over oceans and flown through clouds, landed on Mars and the Moon. Everything humans have touched, fire has touched as well.

Yet it remains a curious technology, just as it was for the Ancients an odd "element." In one form, it is a tool that behaves like other tools. It can apply heat the way an ax can apply impact. A candle holds flame the way a handle holds an axhead. Yet in other forms, it more resembles a domesticated species. It must be birthed, tended, trained; it compels people to change their own habits to accommodate its; it derives its power from its surroundings. Field fires have more in common with oxen than with axes. The hearth fire cannot be put on a shelf as a hammer can. It is more akin to a draft horse that needs a barn, feed, currying, and bridle. There is still one more form of fire technology, and that is fire's status as a captured ecological process that people can, broadly, harness. Humans can tap into the power of air and water to turn gears and millstones, but we cannot call forth floods or gusts in the way we can flame. In brief, fire roams across a wide spectrum of human tinkerings. Moreover, fire is perhaps the ultimate interactive technology because it makes possible other tools. Even where fire does not dominate—might almost seem absent—somewhere along the chain of technologies it serves as a catalyst or enabling device that allows events to proceed, without which a link or two would break.

Its variants do matter, however. To the extent that fire is a simple tool, it is possible for another tool to replace it. An acetylene torch can substitute for a forge, an incandescent wire for an oil lamp. This process has so progressed that the industrial world has little use for open flame, which it regards as impossibly dangerous. Much as early life incorporated oxygen into the machinery of the cell, constraining it to single, well-controlled acts, so modern technology has absorbed fire, until combustion has replaced fire altogether, and concentrated heat has replaced combustion. It is harder to substitute for fire as a kind of domesticated creature, because burning is essential to the task. Fire does a variety of things, not easily emulated in its ecological effects. But to the extent that it burns in a built setting (even one "built" of natural materials), it is possible to reconstruct that setting, piece by piece, with surrogates at each point. This, for example, is the logic of industrial farming. When fire serves its purposes as a loosely controlled ecological process, however, no exchange is possible. What is needed is fire as it freely burns in a roughly natural context. The ability to start and stop this process is surely a technology, but it is not a "tool" as commonly understood. One can break a campfire into its constituent parts, can find alternative sources of heat, light, attraction. One cannot so break down a fire sweeping through a pine forest. The range of its interactions with its surroundings is too complex. To speak of such fires as "tools," as though chain saws and tractors and ammonia fertilizers could substitute, is to miss the point of their presence.

Its titles are thus important. Treating the tamed fire as though it were a mechanical device can cause trouble. It is a truism that how people perceive fire will influence how they respond to its powers and problems. Such perceptions are also complicated, for fire's symbolic power has always matched its practical powers. The care of fire became the paradigm for domestication. The application of fire became equally the paradigm of technics, of the innumerable crafts that require fire or rely on the tools that fire renders and assists. Fire remains, above all, the great transmuter. It is, for poets and philosophers as much as for engineers, the essence and model of change, not solely for the things it personally combusts but for the infinity of things its applied heat softens, melts, molds, speeds up, and powers.

Over millennia, fire has itself been transmuted. No Paleolithic hunter would likely recognize the fire in a pump-action shotgun; no Neolithic swiddener, the flames buried in a tractor or the nitrogenous fertilizer sprayed by a portable power pump; no priest, the theophanous fire behind a fluorescent lamp; no natural philosopher, the fiery prime mover fed on fossil fuel, turning the geared wheels of industry; no poet, the quintessential combustion that makes software possible. In truth, as the third millennium dawns, one can little improve on the observation of Pliny the Elder, the great Roman naturalist of the first century A.D., as he pondered the role of controlled fire on remaking rock.

> At the conclusion of our survey of the ways in which human intelligence calls art to its aid in counterfeiting nature, we cannot but marvel at the fact that fire is necessary for almost every operation. It takes the sands of the earth and melts them, now into glass, now into silver, or minium or one or other lead, or some substance useful to the painter or physician. By fire minerals are distintegrated and copper produced: in fire is iron born and by fire is it subdued: by fire gold is purified: by fire stones are burned for the binding together of the walls of houses.... Fire is the immeasurable, uncontrollable element, concerning which it is hard to say whether it consumes more or produces more.*

Prometheus Unchained

Call them, collectively, *pyrotechnologies.* Begin, however, with the technology of fire itself because the power fire promised could happen only if one could create and control it at will. Fire had to be present when needed and had to exist in a usable form. This required devices to start fire, special fuels to stoke it, and appliances to store and regulate it. They are among the most ancient of technologies and the most familiar, or were until industrialization rendered them alien, almost magical.

Fire Starters, Fire Preservers

Nature has not been an easy source for fire, however, since some places have little flame, and others have it only as the whim and seasonality of lightning or volcanic eruption allow. Nor, for early hominids, was fire easy to make. They had to hold on to it once they had it. If they lost it, they could get more only by begging, borrowing, or stealing from others. Yet it was rare for groups to give fire away. It was too precious. They shared only within a clan, from a common source, and shared with outsiders only during core ceremonies like marriage or treaty-signings,

*Pliny, quoted in Cyril Stanley Smith and Martha Teach Gnudi, trans. and eds., *The Pirotechnia of Vannoccio Biringuccio* (Cambridge: MIT Press, 1966; New York: Dover, 1990, reprint), p. xxvii.

where the commingling of fires symbolized the merging of their interests. To lose fire could be disastrous, the very symbol of catastrophe.

So they strove to preserve fire. Slow matches, banked coals, embers insulated with banana leaves or birch bark, perpetually maintained communal hearths—all kept fire constantly alive. With suitable kindling and coaxing, new fires could be ignited from this source. The effort to preserve the hearth fire or the sacred fire of the larger community had thus an immensely practical purpose, eventually coded in elaborate ceremony and symbolism. Many peoples, moreover, carried their glowing fires with them when they traveled. It was first believed that Australian Aborigines, Tasmanians, and Andaman Islanders, for example, did not know how to start fire because for long decades they were never seen to kindle one. Instead they carried their firesticks with them.

They were right. Fire was usable only if it was portable. Most groups, however, substituted fire starters for fire itself. Three kinds of devices prevailed—the fire drill, the fire piston, and the fire striker. The first includes fire plows and saws, as well as drills proper, which work by vigorous rubbing to the point that the heat of friction can kindle tinder. The second type, more restricted, works like a diesel engine by quickly plunging a tinder-draped piston into a small chamber and then pulling it out. The rapid buildup of heat and sudden release into oxygen results in ignition. The fire striker embraces a wide variety of instruments that shower sparks onto tinder. Drills and strikers closely mimic the stone and bone tools of *Homo sapiens,* and almost certainly date from the same Paleolithic epoch; their geography tracks a map of human migrations.

FIGURE 9. Cooking the woods. Biomass had value beyond its contribution to swidden and pasturage. Even so, people typically relied on controlled heating to distill the essences they desired. This picture from Denis Diderot's famous *L'Encyclopédie* shows how extensive an operation like charcoaling could be. The wood was split into a standard size, then stacked and covered with dirt. Small vents and constant tending controlled the rate of slow distillation; the object was to pyrolyze the volatiles that support flame. The final outcome is an ideal fuel, suitable for glowing combustion, that can produce a steady heat without the pulsings of a flaming front. Still, charcoal is bulky to transport and can quickly deforest a site. As industrial demand increased, fossil biomass replaced it. What charcoaling removed as troublesome volatiles, other techniques sought to capture. Pitchy pine flakes could be "smelted" down into tar and other "naval stores" like turpentine. Variously heating different woods could yield a wide variety of useful, raw chemicals. (Source: Gillispie 1987)

fig. 4

To coax fire from wood or flint must have seemed like the deepest conjuring. Certainly, the ability to call fire forth on command signaled a revolution in fire history.

Over time certain fire starters triumphed, almost to the point of becoming universal. Conquerors and colonizers imposed their own devices; trade bolstered others; Europeans, in particular, promoted the strike-a-light, favored since Neolithic times. (Even the 5,000-year-old "ice man" recovered from a glacier in the Ötztaler Alps had one, along with a pouch for tinder.) Eventually pyrite and flint gave way to steel and flint and joined European traders, missionaries, soldiers, and colonists as they tramped around the world. The technics were, after all, the same as that exploited for flintlock rifles. Then a chemical revolution replaced the awkward strike of steel with the smooth friction of the match. The first (the sulfur-reeking "lucifer") appeared in 1827, succeeded by a phosphorus version in 1830, and the safety match in 1852. No longer did fire starting require either cost or skill. Anyone could call it forth. The ancient bonds of fire tending and codes of fire-related behavior disappeared into pants pockets. But by then, other than for smoking tobacco, there was little reason to haul it on one's person.

Fuels: The Great Chain of Fire's Being

What mattered, though, was preservation, not ignition. A spark was only as robust as its tinder. One solution was to store kindling in pouches. Another was to combine fuel and flame in a slow match or a firestick. As one torch burned out, another would be kindled from it. The firestick could then transfer flame to a campfire, perhaps sheltered, from which another firestick could be wrested when the time came to move on. The flame became constant. The role of fire keeper was essentially that of fuel provider.

Whether closed or open, a tended fire was really a fire well fed. The search for combustibles was endless and often time-consuming. It frequently extended over the countryside, and was a consideration in the periodic relocation of villages. Most settled, agricultural places had to

---

FIGURE 10. Cooking stone. Almost all aspects of mining appealed to fire, as illustrated in Agricola's 1546 treatise. (Top) A sampling of fire used to smelter and refine ore by open-roasting, melting, and otherwise distilling. (Bottom) A miner uses burning faggots to heat and help fracture rock, a cumbersome but useful technique before gunpowder. (Source: Agricola 1546, 1950)

A
D
D
B
C
A

B
B
A
C

grow their fuels, which they did by coppice or the use of stubble or by reliance on the dried dung of their livestock. Regardless of where they got it, they had to stockpile it, keep it dry, split it into suitable forms. It was hard to say which most controlled the other—the fire or the fire tender.

The need for fuel prompted its own technologies. Not surprisingly, most relied on fire—fire-killed forest, fire-pruned coppice, fire-distilled wood such that fire created the fuel for more fire. Perhaps the best known practice involves charcoal, a twice-cooked substance, once without oxygen, once with it. The slow heating of wood in a sealed dome leaches out by pyrolysis the volatiles that encourage flaming. The solids that remain will then burn steadily through conduction, glowing with a steady heat, rather than flaming wildly.

Still, fire could burn everything people brought to it. It could quickly exhaust, if people chose, whole countrysides. The lust for more fire—checked only by the ability of surrounding landscapes to grow biomass and people to convert it into combustibles—eventually led to an unbounded fuel source, fossil biomass. Fossil fuels existed as coals, lignites, oil shales, natural gas, petroleum. The latter, in particular, inspired its own pyrotechnology for chemical distillation, which made it also immensely portable and vastly more potent. But refined fuels required refined combustion chambers. Automobiles could not run on wood or coal; refrigerators and heat pumps could not function easily with furnaces; power lawnmowers could not survive on steam. The creation of new fuels, in brief, not only made possible but demanded new fire appliances, new tinder pouches, new hearths. The fusion of fossil fuels with fire engines, each rapidly redesigning the other, traces the fast spiral of industrial fire.

Fire Appliances: Creating Specialty Habitats for Fire

The place where spark and fuel met decided the traits of the domesticated fire. Fire proved enormously malleable—flame had no fixed form, firelight no necessary brillance, and the heat of combustion no inevitable flow. All could be molded, and over time each property was selected much as dogs and horses were bred for size, speed, coloration, and sense of smell. The chosen means was the combustion chamber, which controlled the movement not only of heat and fuel but of air. And more: honing the fire required that air be refined into oxygen and rough biomass into its chemically active parts. What oxygen was to air, this distilled combustion was to fire.

FIGURE 11. Fire and sword. Fire has a long association with battlefields. Increasingly, concentrated fire became a potent weapon. (Above) The cannon replaced open flame and smoke with the modern firepower of explosive powder. (Below) An arsenal at the incendiaries ready to be launched against cities, particularly their vulnerable roofs. (Source: Biringuccio, 1540, 1990)

Until recently, however, these contrived keepers of specialty flames still put fire before its human tenders in a very direct way. Fire's presence was undeniable, however it might be encased in brick or metal or sited above tallow or pipe. But industrialization has changed that. Flame no longer appears before people, or for that matter before nature, in a visible way. Technology has progressively separated combustion from flame and segregated the chambers where burning occurs from the places where its energy is felt. No one cooks over a dynamo. Electricity has erected a firewall between source and sink greater than any masonry bulkhead. Fire exists covertly in its products rather than overtly by its active presence. It flourishes subliminally in the cement, brick, tile, glass, silicon wafers, metal, incandescent lights, refrigerators, heat pumps, and gas-propelled vehicles that populate the modern world. Industrial appliances have done for the evolution of natural fire what genetic engineering promises to do for the evolution of life.

So, too, industrial fire rarely meets directly with the biological Earth. Combustion occurs outside the biosphere and within mechanical casings that have so divided burning into its constituent reactions that the outcome qualifies only minimally as fire. Controlled flame rarely strikes trees, soils, or scrub, or the creatures that live amid them. It encounters fire through its servant machines. Yet this is sufficient for industrial combustion to fundamentally restructure the ecology of fire on Earth. The modern Earth's flow of matter, energy, and organisms increasingly follows the stream of industrial combustion. Even the Earth's climate teeters on a geologic tightrope as long-buried biomass, passed through the pyrotechnic flames, bursts forth into its atmosphere, layers its continents, and sinks into its oceans. No true flame could do more.

How Fire Fights Fire

Controlled fire has come full circle. Its first seizure led to a program of captive breeding that ended with fire crumbling into chemical shards. The once-visible fire is becoming a virtual one. Preindustrial fire could always, if it escaped, revert to type, leave hearth or forge and become feral. Industrial fire cannot. Pyrotechnologies have refined the hearth fire to the point of extinction. Still, the process of replacement does not stop at the hearth—is not content to merely displace open burning—but has pursued flame wherever it appears. It has sought to remove all free-burning fire, indirectly by substituting for it, directly by suppressing it.

From the beginning, controlled fire has been humanity's primary means of containing wildfire. People protectively burned fuelbreaks and patches to retard fire's spread, and countered wildfire with backfire. But now modern technology has removed fire even from firefighting. For industrial countries, fighting fires has ceased to mean the clash of one flame against another; now free-burning fire is suppressed by using the engines and preburned bricks and cement of industrial combustion. Two fires cannot, it seems, both claim the same niche. If a new species of burning arrives, it somehow means the old ones must depart.

Cycles of Pyrotechnology: How Fire Has Cooked the Earth

Just as fire turns the gears of ecological cycles, so it has cranked the cycles of many of the things people do to make that ecosystem habitable. Consider three examples, all of them variants of cooking: the cooking of food, the "cooking" of rough biomass, and the "cooking" of rock. For each, fire is the great enabler. Remove it and the cycle collapses.

Cooking as Pyrotechnic Paradigm

In many fire-origin myths, a protohumanity laments as a cruel hardship that it must eat food cold or raw and has no means of preserving food other than by drying it in the sun. The capture of fire changed all this. Cooking became the very emblem of the domesticated fire. Out of the campfire and hearth arose the kiln, the furnace, the forge, the crucible, the oven, and the metal-encased combustion chamber. From cooking food it was a short step to cooking other matter—stone, wood, clay, ore, metal, the air, even seawater, whatever fire could transmute into forms more usable to people. In effect, humanity began to cook the Earth.

Cooking was, in fact, only one phase in a long-wave cycle of food preparation for which people might resort to fire at nearly every stage. Fire helped pluck or massage the food out of the larger biota; fire cooking followed fire hunting, fire foraging, fire-based farming and herding. Fire helped ready meat, grain, or tubers for eating, improving the taste, leaching away toxins, killing parasites. Fire—its heat, its smoke—then helped preserve what was not instantly eaten.

It is difficult today to comprehend how pervasively fire could affect this process. But as an interesting illustration consider how pyrotechnology shaped the economy of food for 16th-century American Indians as recorded in Thomas Harriot's *A Briefe and True Report of the New Found Land of Virginia.* To paraphrase: The axis of the village passes through a great fire, around which the tribe stages its "solemn feasts." The hunting grounds for deer they keep open by regular burning, and the deer themselves may also be fire-driven into streams or coastal tidewaters during a fall hunt. The crops of maize are swiddened. The houses have hearth fires. But, unexpectedly, the cycle extends even to fishing. With fire the Indians fell trees and hollow them into boats. They carry fire in the craft while they spear for their prey and at night the torch draws fish toward them. They broil their catch over flames, or cook it in an earthen pot along with maize and other foodstuffs. Any surplus fish they dry and preserve, also over fire and its trapped smoke. After the meal they celebrate or offer prayers around a "great fyer." Like their village, their lives and their economy are centered around a flame.*

Cooking Woods

If fish and venison, maize and cassava could be cooked, why could fire not "cook" the landscape for other goods as well? Ancient chemistry was largely cooking applied to other substances. Whether the change sought was physical (a change of state) or chemical (a change of substance), fire wrought it. Fire could break apart, distill, soften, stiffen, encrust, melt, or transmute. But by way of example, consider how people could cook the boreal forest of northern Europe to feed their general economy.

The range of things heated, steamed, boiled, or roasted is huge. Of course, there was widespread swidden farming, without which cultivation was impossible, and broadcast burning for pasturage, essential to livestock and especially dairy products. Beyond that, however, it was possible to chop up and cook the remaining forest for human purposes. One could collect and open-burn the unfarmed woods (aspen was particularly desirable) to get potash, a valued source of potassium used as fertilizer in farming and in the manufacture of goods from soap to gunpowder. One could anaerobically burn hardwoods to get charcoal. One could slow-cook pine to siphon off tar, pitch, turpentine, and other fugitive distillates that made up the "naval stores" industry (so called

*Thomas Harriot, *A Briefe and True Report of the New Found Land of Virginia* (New York: Dover, 1972, reprint of 1590 edition), pp. 69, 55, 56–57, 60, 63, 66.

because the products were vital to wooden ships). Scoring patches of pines—a kind of raw orchard—assured a good supply of pitch as the trees poured forth sap, which then hardened, to cover the injuries.

Through such means people could colonize an otherwise uninhabitable forest, one often plopped on morainic soil resistant to the plow and in a climate hostile to winter grazing. What foods people could not cultivate locally, they could trade for. That traffic, of course, relied on wooden ships, which got their masts from the Scots pines that sprouted in dense throngs in the aftermath of fires, their caulking from the tars and pitch distilled from lesser pines, and even their ropes from the hemp that flourished on burned plots. Little of the landscape escaped. Its human residents bent such places to their will with a kind of second-order firestick farming, sometimes on the scale of individual trees slashed for pitch or tapped for resin, often of swidden-sized patches cultivating charcoal or potash. Without fire to rework the woods, however, their labor meant nothing. Without their fires they were little better off than moose or voles.

Cooking Stones

The firing of rock is perhaps more spectacular because it has no obvious natural origins, save perhaps volcanoes. (The Roman philosopher Lucretius thought that a forest fire had led to the discovery of metallurgy by melting outcrops which then dripped copper and iron, but most readers parse those passages as poetic license.) The more likely inspiration was cooking. Miners roasted ore as they might pork, boiled down liquids as they did syrups, poured molten glass and iron as they might jelly. A mining complex resembled nothing so much as a vast industrial kitchen.

Preindustrial mining exploited fire at every turn. Prospectors burned over hillsides to expose rock. Miners relied on fire to tunnel, to smelt, to forge. Only the very richest and nearest mines could afford to haul raw ore very far. They had to crush and process as much as possible on site, and nearly every stage demanded fire. Accordingly, mines were only as good as their fuel supply, which until recently meant wood or charcoal. The great copper mines of Cyprus, for example, grew, cut, and regrew the surrounding pine forests a score of times over the centuries; the Rio Tinto mines in Spain engorged 42 tons of wood a day, amounting to 3.2 million hectares of woodlands over its lifetime. The origins of forestry in Sweden and Russia lay in the state's desire to promote the fuel-laden woods around great iron mines.

Within the mining cycle, fire figures repeatedly. Georgius Agricola's great treatise, *De Re Metallica* (1556), is a grand introduction, cataloguing practices that date from ancient times to the onset of the industrial era. Where the veins resisted their iron picks, hardrock miners lit fires to shatter the stone sufficiently to pry out ore. This was dangerous work, requiring that mines consider ventilation; but miners already relied on fire to illuminate the shafts, and it was only a matter of degree to put their torches to the stone directly. Eventually gunpowder replaced wood and steam. Yet "fire in the hole" endured.

With fire, assayers tested the ore to determine its character and value. With larger furnaces or pyres, some open, some enclosed, they roasted and cooked crushed ore. The process varied with the properties of the metals involved, the abundance of fuels, and local traditions. But at some point all metallic ores would be heated either to separate them from their embedded rock mass or to liquify them so they could be poured and shaped; most often both. Hotter fires required a special chamber, proper fuels (at a minimum, charcoal), and control of the air supply, preferably by means of a bellows. Eventually the furance became a forge to further refine and mold.

But nonmetallic stones often demanded firing as well. Turn to Vannoccio Biringuccio's *Pirotechnia*, published in 1540, for a splendid survey of fire's pervasive presence in every metallic (and any other) mining that involved chemical changes. Limestone could be roasted into calcinated lime suitable for cement, sand melted into glass, clay baked into ceramics. Sulfur, mercury, and alum all depended on chemical fire to pluck them loose from gangue and then to purify them into their elemental core. Then there are the distillates: salt from seawater, nitric acid from *aqua fortis,* alcohol, oils, and "sublimates" in general. Almost any chemical reaction—the "art of alchemy," whether true in its larger claims or not, thought Biringuccio—relied "on the actions and virtues of fires." Fire was the chemical fulcrum by which humanity could leverage even its mechanical power, by which it could make and move the hard tools that together reshaped first-world nature into a second world of human contrivance. (More ominously, he concluded his treatise with fire weaponry, cataloguing devices that rely on fire to hurl projectiles or on the projectiles to kindle fire.) In the end, the lithic cycle feeds itself: the iron burned out of the earth becomes the picks and shovels by which miners can dig more ore and the axes by which to cut the timber they require for shoring and—most ardently—the fuel they need for smelting and forging.*

The cycle returns, back on itself. While Biringuccio concluded with an extended metaphor on "the fire that consumes without leaving ashes, that is more powerful than all other fires, and that has as its smith the great son of Venus," the fires of *Pirotechnia* needed something real to burn. Here biomass had an advantage: it could be more easily cooked because it could itself burn. Stone could not; it took heat, but didn't give it. It continued to only absorb until industry found ways to burn fossil biomass. What had once seemed an absurdity, the self-combustion of rock, has in fact become the basis for our modern pyrocivilization.

Fire Powers: Controlled—and Not-So-Controlled—Fire as Mover and Shaker

Burning trees for ash and pitch could appeal to nature for its inspiration; burning stone less so. But in both cases fire set by human hands met natural objects. There is no intrinsic reason, however, why humans had to restrict their torches to what nature presented. Nor did they: pyrotechnology could go where people pleased and could obey just as readily logics other than those proposed by nature.

Consider in particular warfare and engines, whose dynamics derived from politics and economics rather than wet-dry cycles and the pyric chemistry of living biomass. Their ecological impact was sometimes overt, as when battles set fires that roamed across fields and woods. More often their ecological clout was disguised, an iron fist hidden in a velvet glove of economics. Fire weapons and fire engines restructured the flow of goods and peoples, influenced how people used the land, and quickened the tempo of technological change. They rearranged fuels and invented new fire devices. They plunged whole landscapes into a forge of human fury and ambition.

"They Laid Waste and Burned": Considering War as Fire Ecology

War has long been associated with fire. "Fire and sword" very nearly says it all: open fire, as a tactic of battle, as the scorched earth of retreating armies, as the laying waste by victors; closed fire, as the means of

*Smith and Gnudi, trans. and eds., *Pirotechnia*, p. 336.

forging weapons, of casting cannons, of powering ordnance. "Firepower" remains the codeword for military strength.

Few battlefields have lacked fire. Fires have burned on prairie and woods, amid ships and cities, flung over ramparts and scattered with artillery shells. Fire weapons have traveled on land, sea, ice, and in air. Yet open fire could be problematic, and nowhere more than amid the havoc of battle. Clauswitz's "fog of battle" was most often a cloud of smoke. A broadcast burn could, with a change of wind, turn on those who set it; smoke screens obscured the field for both sides. Even in naval battles, the ideal was to hurl enough controlled fire to disable a wooden ship, not enough to destroy it. Sieges sought to burn out defenders, while soldiers on the battlements poured down flame on assault troops. In the ancient world, Greek fire (a sulfurous liquid) was a weapon to dread. For gardened societies, especially, the chaos of war invited the chaos of wildfire, since the breakdown in social order exposed niches for fire and strewed the landscape with fuel.

Over the past millennium two revolutions in firepower have shaken the conduct of war. One was gunpowder (which gave new meaning to the expression "to fire"), and the other, industrialization, which mechanized war and expanded its range. While each fabricated a host of new fire weapons, it is often easy to miss the flames for the roar. The worst casualties of World War II resulted from blasting cities with a mix of "conventional" blockbuster bombs and incendiaries. Even the atomic bombs used on Hiroshima and Nagasaki did their greatest damage through the fires they kindled. The U.S. Strategic Bombing Survey concluded that four-fifths of the destruction wrought on British and German cities by aerial bombardment was "fire damage," that "incendiaries, ton for ton as compared to high explosive bombs, were approximately five times as effective in causing damage," and that the aerial assaults on Japan were "frankly fire attacks." If fire seems increasingly invisible on modern battlefields, it is because the flames have vanished into tank engines, cartridges, and rockets. But even the Gulf War, fought on incombustible sands, ended with burning oil fields. It was, after all, another fire war, fought over the fuels of modern industry. Perhaps not so oddly as it seems at first, those flames will likely endure as the unquenchable symbol of that conflict.*

*Percy Bugbee, "Foreword," in Horatio Bond, ed., *Fire and the Air War* (Boston: National Fire Protection Association, 1946).

As that black pall, spreading over the sky like an oil slick, shows, waging war with fire has ecological effects. For some landscapes—temperate shade forests, mangrove swamps, cities—war-hurled fire is a major disturbance. Battlefields are shaken landscapes; fire ordnance is a great slasher-and-burner of towns and forests. A little weirdly, this is not always ecologically evil. Training fields in East Germany churned by tanks and shells led, after unification, to nature reserves of exceptional biodiversity. Mostly, though, the biological impacts of military fire are muted and hidden, as are other forms of industrial combustion. War quickens the pace of technological development, redefines and sometimes replaces societies and their economies, realigns politics—all of which can break and burn landscapes as thoroughly as any conflagration.

The Power Within: How Fire Engines Became Prime Movers

Still, the more revolutionary fire is that encased in metal and used to power pistons. With the steam engine, the stationary fire became more than a hearth-evolved furnace: it apotheosized into a prime mover. The fast combustion of fire engines could compete directly with the push and pull of slow-combustion muscle. As Matthew Boulton, James Watt's partner in combustion, succinctly told a visitor, "I sell here, Sir, what all the world desires to have—power."*

The problem, as so often, was fuel. The steam engine could not by itself break down the ancient ecology that bonded burning to biomass. The early engines were furnaces, not unlike distillation systems, except that the boiled-off steam could drive a piston. They burned cordwood (or charcoal), which left combustion ultimately at the mercy of what the countryside could grow and operators could glean from it. Consuming staggering quantities of wood, they could rapidly burn up whole landscapes. That set in motion the search for a more robust fuel, a quest that ended with coal.

Fossil fuels had long been burned, but locally and specifically because they lacked a place in which to combust usefully. They could not be spread over fields like branches, or rolled like smoldering logs, or loosed as flame could over once-living fallow. The steam engine thus gave coal what it most lacked, a combustion context. In return, coal granted the new fire engines abundant fuel. They soon worked on one another, coal

*Boulton quoted by James Boswell, *Life of Samuel Johnson* (22 March 1776) (London: G. Bell, 1884).

encouraging better designs, engines seeking more refined fossil fuels. Together they revolutionized power machinery and transport, and through transport, all the landscapes internal fire could touch. The steam engine soon spawned other combustion-driven prime movers that could burn more portable fossil fuels like petroleum and natural gas. Each innovation bred others. Eventually this swarm of fire-breathing machines forced fire ecology into another order of being. They made possible industrial fire.

Even oblique means can sometimes yield awesome ends. That is what steam did to fire. Combustion no longer flowed from living source to living sinks. It burned biomass from the geologic past and released its outflow to a future Earth. The ancient chain of combustion no longer resembled anything in its past. Industrial combustion substituted its closed fires for open ones and attacked free-burning fires seemingly wherever it found them. And it relocated fire ecologically by breaking down and rearranging landscapes, helping decide what might burn and when it should burn and by what means. Although a robust ecological understanding of industrial combustion still eludes us, through its engines Third Fire has become the prime mover of Earth's fire regimes.

FIRE IN THE MIND

Fire lived in the mind as well as on the land. It had to be explained. It loomed too large in human experience not to cry out for a story, a theory, a personification. It became a source of myth that explained how and why humanity differed from the rest of creation. It appeared as a deity, whether as wrathful smoke and flame on Sinai, the mischievous Loki or unpredictable Agni, or Vesta's gentle glow in the hearth. And it puzzled natural philosophers for long centuries. It was one of the four basic elements for ancient Europeans, one of five for the Chinese. But it was not truly an element, rather a reaction so basic that it seemed elemental. Heracleitus announced that all things were an exchange for fire and fire for all things. Eventually fire became more powerful as a means by which to explain what happened in the world than as an object to be itself explained.

Flame became a mental tool as well as a practical one. It was the essence of change, especially rapid change. Just as people used fire to remake the world around them, so, philosophers reasoned, must nature. It was a simple step to argue that fire, which shaped so much of the world, also shaped the larger universe. Philosophers instinctively turned to fire as much as cooks did, and experimental science appealed to fire as technologists did. Besides, flame fascinated. Even the Enlightenment stared hypnotically. Philosophes *were as convinced as Pliny that fire was everywhere. Earth had its central fire, the solar system its solar fire, and the heavens the celestial fire of the stars, comets, and quintessential aether. Electrical fire discharged as lightning. Inner fire provided the life force for plants and animals, the source of animal heat. And of course there was the ever-enthralling fire in the machine. In 1720, Hermann Boerhaave confirmed the supremacy of fire by declaring that "if you make a mistake in your exposition of the Nature of Fire, your error will spread to all the branches of physics, and this is because, in all natural production, Fire ... is always the chief agent." Even as late as 1848, when Michael Faraday wished to demonstrate the principles of natural philosophy, he chose, on ancient precedent, fire for his subject.**

*Boerhaave quoted in Gaston Bachelard, *The Psychoanalysis of Fire,* trans. Alan C. M. Ross (Boston: Beacon Press, 1964), p. 60.

Yet Faraday's Chemical History of a Candle *also helped complete the intellectual transmutation of fire, its collapse from a universal cause to a chemical consequence, the mere motion of molecules, the quantum bonding of oxygen. The transition occurred—not accidentally—with fire's condemnation by agronomists and foresters, with its removal as a vital force in urban life, and therefore in the felt life of the educated elite who lived there. Gaston Bachelard might boast that he "would rather fail to teach a good philosophy lesson than fail to light my morning fire," but most philosophers no longer lit fires or cared to understand them. The American Ben Franklin, for example, tamed "electrical fire" through his lightning rod, caged the wasteful hearth fire inside a metal stove, and devoted his philosopher's mind to electricity rather than the elemental flame.**

That, in brief, is what happened across Western civilization. Technology provided the models for nature, instead of nature for technology. Industry invented new pyrotechnologies, then suggested that heat engines were an analogue for animal heat. Natural philosophy found other ways than fire to explain the world, and then used that revealed world to explain fire. Chemistry downgraded flame to an atomic reaction. Thermodynamics split fire from motion and heat, electromagnetic theory from light. Fire shrank from Heracleitean universality to a laboratory demonstration. Once the manifestation of the deity and the source of life, fire had become alien, a destroyer of cities, a savager of soil, a befouler of air, an emblem (in science as in agriculture) of the hopelessly primitive. Long an informing metaphor, philosophical fire became a cliché, fit only for humanist scholars and the garish covers of romance novels.

By the time ecologists realized that flame had a vital role in many biotas, they had as little intellectual heritage to draw upon as they had practical experience. The fires that had once surrounded humans and illuminated and shaped their world no longer existed for those societies that had elevated the study of nature into modern science. The more sophisticated the scientific culture, the more likely the closed combustion of Third Fire had squashed or confined the open flames of First and Second Fires. Modern fire keepers would have to rekindle ideas out of new tinder.

*Ibid., p. 9.

Chapter Eight

Frontiers of Fire (Part 3)

FIRE COLONIZING BY EUROPE

It helps to remember that the geographic expansion of Europe resembled that of other peoples. The slow saturation of continental Europe by sedentary farmers matches the southward migration of the Han Chinese, both of them crowding swiddeners and herders to the margins. The long reach eastward across Eurasia by Slavic peoples echoes the great probes of Bantu speakers southward through Africa. Even the expansion's seaborne phase recalls the Austronesian diaspora, which was also committed to remaking lands according to the precepts of agriculture. That Europeans moved plants, animals, diseases, and peoples beyond their ecological hearths—sometimes far from their places of origin—had ample precedent.

But this expansion differed in scale, the shock-intensity of the encounter, and the extent to which Europeans plunged on until they reached every hill and stream on the planet. Even those distinctions, however, pale before the venture's catalytic power. At its midpoint, Europe industrialized, and Europe's imperial outreach became the vector for spreading Third Fire over the Earth. As a fire planet, the Earth looks the way it does because Europe sailed beyond its confining shores and eventually hauled the industrial revolution under its sails.

No previous diaspora had the sheer global sweep of Europe's. What Europe did not colonize outright, it affected indirectly through political meddling or commercial contacts. Some landscapes, like Australia, were simply overrun with European colonists, but many more adjusted their land usage and fire regimes to the European presence. Everywhere Europeans observed such changes, but not everywhere did they like what they saw. Too often contact meant a kind of ecological plundering—culling the most valued trees, the bulk killing of fur seals, dodos, and passenger pigeons. A landscape appeared of eroding soils and drying springs, of biotas beaten down and infested with weeds and pests. And of course everywhere those Europeans saw fires—strange fires, wild fires, devouring fires.

Such observations did not mean much, however, until the scientific revolution outfitted European thought with both the means to assess the change and an apparently rational program by which to correct it. The Enlightenment could measure, and it could criticize, and it did both. Moreover, it flourished amid a renewed surge of European exploration and colonization, and proposed a rational reaction. Critics argued for programs of resource *conservation,* which in turn required state-sponsored agencies to oversee them along with a program of scientific study to ensure that they were right. The result was the invention of institutions often global in their geographic sweep and universal in their intellectual assumptions. Those became as much a feature of Europe's ecological imperialism as trading companies, folk migrations, market-driven extinctions, and wasted forests.

Thus it mattered hugely what Europe thought about fire. Since most Enlightenment emissaries came from temperate Europe, flame burned more brightly in the colonies than in the homelands. Whatever happened seemed to happen with fire on hand. Colonists applied it without the social shackles fire practices had known in Europe, and natives without the legitimating context of European cultivation. It was but a small leap of logic to suggest that to control fire was to control the land and its peoples. Indeed, greater coercion was possible overseas than at home. But even as Europe weighed its judgments about what fire was right and proper, it was itself undergoing a revolution in combustion more profound than any since Prometheus handed humanity the torch.

Industrialization combined with imperialism to make, move, and dissolve fire frontiers. The fire geography of the Earth today is largely the outcome of what an imperial and industrial Europe did, or tried to do.

How Europe Expanded Fire's Realm

Even as Europeans marveled at Tasmanian Aborigines who walked everywhere with their firesticks and at Virginia Indians who speared fish with open flames nestled in their canoes, they themselves wore strike-a-lights with their bucklers while their frigates held flame constantly in the hold. Their own fires they hardly noticed. Yet they, no less than the peoples they met, traveled with fire near at hand and used it to make habitable the places they encountered. Above all, fire was the ecological enabler that, rightly used, made European agriculture possible.

Europe's Grand Narrative of discovery and colonization was a story of the torch brought to new lands. Rarely did the clerical classes see it that way, but so it was. The lesser stories of that master narrative were many and varied, as one should expect; yet three may justly serve to illustrate the span of possible plots. One hauled swidden into new lands, one herded livestock beyond their natural range, and one yielded a hybrid of European and native practices.

Finnish Colonization: Making New Lands, Remaking New Worlds

Over the centuries landnam, like the far-wandering tribes that carried it, ceased to roam through temperate Europe and put down roots. Still, eager farmers continued to probe and punch along the borders. Where Slavs and Finns met around the 10th century, an agrarian hybrid resulted—a rye-cultivating swidden that showed extraordinary vigor as a pioneering force. The Slavs moved east, the Finns north. The Finnish surge hollowed out the coniferous interior of the eastern Baltic into a vast, fallowing forest. The system pushed north and curled, improbably, around the Gulf of Bothnia. Meanwhile, the Swedish monarchy, eager to develop its interior estates, imported Finnish swiddeners to repeat the process through central Sweden. (Most ethnic Swedes clung to the coast.) Again, the pioneers pushed out to all sides. In fact, they moved so robustly that within two centuries they found themselves compared to locusts and denounced as pests.

But Sweden's ambitions extended overseas as well as to Norrland. In 1638 it erected a trading colony along the Delaware River in North America. Among those who emigrated, either willingly or under force, were clusters of Finns from prime swidden regions like Dalarna. The colony failed, succeeded by the Dutch and later absorbed by the British. The colonists, however, remained. Between them and the native Lanapi Indians, another fire-tempered hybrid emerged, an ideal vehicle for pioneering.

Swidden, free-ranging herds, long hunts, log cabins, all the trappings of the backwoods frontier crystallized, and then surged west over the Appalachians. But the system's strength was also its weakness. While it was a marvelous device for pushing into new lands, it left to others the tedious task of mopping up, of transforming first-broken woods and meadows into settled farms and fields. As often as not, the same pioneering peoples moved on until they reached the sea-of-grass prairies. Prior to their arrival, much of the temperate woodlands had known agriculture, or had known and lost it through immense pre- and post-Columbian migrations. The swidden survivors of New Sweden restored it.

Transported Fauna: Dreamtime Australia Becomes Domesticated

Among the continents, Australia suffered the most extensive loss of Pleistocene megafauna. Some 86 percent of animals over 44 kilograms died out. The linkage with human colonization is tight: the timing of contact with loss is as close as dating techniques permit. Throughout, climate had remained broadly unchanged. With spear and firestick—hunting and habitat conversion—Aborigines replaced Australia's megafauna with themselves and a rich brood of smaller creatures. Yet a faunal void remained. Humans did not consume all the biomass that the vanished animals had, and unlike other continents, Aboriginal Australia was not farmed. It would not have its "surplus" growth burned for swiddens. Instead it yielded ample fuels for free-burning flame. It was not subject to a faunal recolonization until Europeans arrived and unloaded an ark of animals with eager teeth and (unlike Australia's natives) hard hoofs. The creatures spread like plagues.

The First Fleet, arriving in 1788, transported Britain's mixed agriculture along with its convicts. In their holds the ships carried a floating farm, with 2 bulls, 5 cows, 29 sheep, 19 goats, 74 hogs and sows, 18 turkeys, 35 ducks, 35 geese, 209 chickens, and 5 rabbits. Although field farming proved difficult, herds of cattle began to multiply, and the search for new pasture spurred early exploration across the Blue Mountains. A full-blown assault waited for the introduction of merino sheep, and by the 1840s far-roving flocks sprawled across the landscape of southeastern Australia and soon swirled throughout the interior grasslands. Other livestock followed: more cattle, particularly for the tropical north and for milk herds, along with horses, oxen, and camels. Settlers introduced domesticated pets—cats and dogs. They transplanted foxes. And they unleashed rabbits.

The faunal colonization of Australia was as much a matter of animals that strayed as those that hewed to their flocks, none more so than the European rabbit. The first of them escaped from their warrens outside Geelong in 1859. They bred in the wild and migrated, and by the 1890s invaded every potential ecological nook and cranny, and beyond. Control programs failed, terrain-spanning fences failed, bounties failed and may perversely have aided the rabbits' spread. Not until a viral disease specific to the European rabbit, myxomatosis, was introduced in 1950 was there any hope of real control.

By then several crashes had already buffeted the sea of feral fauna. Rabbits, other livestock, and native wildlife had combined with drought to strip many landscapes of their vegetation, devouring the fuels that would otherwise carry fire. Thus fire regimes were rubbed out as thoroughly as the fire-wielding Aborigines had added them. Though herders often burned—sought the "green pick" that otherwise eluded them—they were too late. The land came back to scrub rather than to grasses. Fire in the bush became more taxing to start, more vexing to control, less predictable in its outcomes. Flame remained, but so altered was its regime that it was rightly seen as new.

Avatars and Hybrids: India Absorbs

Though Europe was loath to admit the fact, its agriculture had limits. Some places were too dry, too barren, too remote. Europe's cultigens withered in the noonday sun, its livestock starved on long-leached soils. Besides, many sites—often the best—already supported a thriving agriculture, and Europe could offer nothing better. But there were also places where, after an initial collision, native practices reconciled with European markets. The politics of compromise often focused on fire, for which the British experience in greater India is a superb example.

Imperial Britain's program for modernization extended to lands as well as to bureaucracy, law, and telegraphs. Britain wanted both a more market-driven agriculture committed to exportable commodities, which argued for clearing and plantations, and more extensive forests, which (its naturalists claimed) could help regulate rivers and stabilize the climate. Officials concluded that traditional burning for swidden and grazing did nothing to advance these ends and likely worsened conditions. Since fire was indispensable for indigenous farming and herding, as well as its most vibrant symbol, colonial rulers concluded they would have to contain the flames. Locally, they succeeded, sometimes all too well. Instead of routine surface fires lightly washing the understory, they got rough, erratic, often lethal burns. More ominously, some of the prime timber species like teak and sal failed to reseed (or if they resprouted, refused to thrive) on unburned sites.

Native swidden was unacceptable, exotic fire protection unworkable.

What slowly grew up in their place was a crossbred system that mixed traditional farming with commercial trees, an early variant of agroforestry. A European precedent existed, for central Europeans had long sown oak or pine seeds into their abandoned swidden plots to ensure themselves of lumber, woody fallow, and tannic acid (from oak bark). The adaptation in Bengal replaced European trees with teak and sal. Officials left to local farmers the exact prescriptions such that sal became intersown with rice, cotton, maize, and sesame. The fire-fallow *Brandwirtschaft* of Europe became the *taungya* of south Asia.

Rather than abolish fire, as it had intended, European agriculture exercised a kind of indirect rule. The old regimes stayed on, with slightly different rhythms and crops. By 1932, *taungya* had become a "universal prescription." One of the foresters who oversaw its evolution, E. O. Shebbeare, concluded that "it must be admitted that our belief in fire is based more on what we feel than on what we know, but the fact remains that aboriginal villagers, who know more than we do, are strongly in favour of burning." Burning persisted, sal and teak flourished, and *taungya* skipped to other imperial, usually tropical, lands where it grew pine, eucalypt, and *Gmelina.**

How Europe Contained Fire's Realm

For Europe's colonizers, starting the fires they wanted was only half their task. The other half was to stop those they didn't want. The usual verdict: our fires are good, theirs bad. That meant, however, not only banishing native peoples' burning but also the fires that escaped from colonists because of carelessness. It was not a long step from condemning native fires for gutting forests and stripping humus to condemning all fires, since they did more or less the same ecological jobs. The strategy suited nicely the instincts of Europe's clerical class, which had never trusted open flame and longed to hound it out of existence.

There were plenty of reasons to worry about what colonizing did to lands. The waste was often mind-boggling, the economic losses staggering. Fire was but one expression, if a worryingly visible one. In response, a political philosophy emerged that became known as conservation because it sought to conserve—to regulate, not eliminate—the basics of land, water, wood, and wildlife. Its practical expressions were institutions

*E. O. Shebbeare, "Sal Taungyas in Bengal," *Empire Forestry Journal* 11 (1932): 25, 30, 32.

like agricultural and forestry bureaus, the gazetting (official establishment) of public lands, and the creation of nature reserves. While not a strictly imperial invention, conservation seemed most essential along the colonial fringe, where landscapes were in greatest upheaval and the authority of the state less compromised. How these schemes played out varied enormously, of course, but the big divide was between those lands colonized through a demographic takeover by emigrant Europeans and those in which Europeans ruled over a subject, usually sullen, native population. That determined who gained and who lost and who held the torch.

Conservation: The Politics of Damage Control

Conservation was an old idea in new bottles. It updated classic European traditions of land use with the authority of modern science and the modern state. Its conceptual core held an agricultural vision, that one could harvest only what one had grown. Merely hauling the output away was nothing but looting the land, and that sort of biotic plundering could not long continue. Conservation updated that agronomic model to incude other features of nature's economy such as forests. In doing so, it also proposed a scheme for thinking about how (if at all) fire might belong and should behave.

Europe's ancient legacy of agriculture declared forcefully how difficult it was to improve yields. One closed cycle led to another, without escape. More fertilizer could improve crops, but fertilizer was, ultimately, grown in the form of dung or fallow. Improved plowing and weeding could fatten output, but draft animals and laborers had to be fed, and those needs could easily wipe out the gains. More land under plow brought more feed, but expanding arable or sown pasture had its costs, particularly when only more marginal and infertile lands remained. And so it went: each gain brought an equal loss.

Nothing outraged agronomists as much as fallow. Here was land unused, and its weedy "waste" only went to feed the flames. Even in the best rotation, a third of the land remained under fallow. A fire-fallow agriculture could never break out of its biotic bondage or stretch cultivation beyond its natural geographic limits. Nature's economy could never boom if it burned up its surplus growth instead of reinvesting it in the soil. Officials, scholars, and agronomists all agreed then that burning was primitive and irrational, that the more fire swept a land the less productive that land must be.

Enlightenment science challenged farming's inherited wisdom while it sharpened beliefs about burning. Particularly the agricultural revolution (which preceded the industrial) showed how systematic experimentation in breeding and cultivation, using new crops and legumes, and adopting redesigned plows and new attitudes—quashing peasant tradition and its superstitions—could fatten yields. In short, Enlightenment agronomy promised that Reason could pull farming out of its sloven fire-fallow ruts, putting the fallow to productive use and shrinking open flame to a trivial role. Less waste, more yield, less folklore, more science, all with the political heft of an Enlightened state behind it—that was conservation.

Colonizing afforded ample evidence of what the reckless, the greedy, and the ignorant could do, and why conservation was necessary to keep them from doing it. Even as Europe's powers grew, so did its capacity to observe and ponder. Most thoughtful observers agreed that Europe, at least by the 19th century, had smashed as many native landscapes as it had native armies. For them colonization had become a kind of semi-controlled experiment. Contact served as a cause, and they (especially the naturalists) recorded the effects. The scale of havoc appalled them. Moreover, contact, particularly the clearing of forests, seemed to pull in its train a tiresome cycle of drought, flood, hardship, fuel shortages, and famine. Plants shriveled on the stalk even as rivers spilled over their banks, and having fired everything imaginable residents found themselves short of fuel to burn for household needs. Surely, principles of conservation should apply. And since colonization was a state-sponsored (or at least state-encouraged) endeavor, conservation also deserved the attention of the state.

But there was also the waste wrought by native peoples whose land use was often more primitive, by European standards, than that of European peasants and even more reliant on burning. Wherever they turned, critics saw flame, smoke, ash. They saw it in the slash of reckless logging and of woodlands cleared for plantations, in the hunting grounds of aboriginal peoples and the seasonal pastures of migratory herders, and in the swiddens of wandering farmers, the deep-dappled landscapes of fallow, both those hewn by natives and those freshly hacked by immigrant pioneers. This crisis, too, demanded the learned power of the state. No other presence could intervene between village and global market, and no other authority could command obedience, at gunpoint if necessary.

So as conservation matured, it blossomed into a philosophy, a program of action, an agency of government oversight, and a scheme of inquiry. Its natural allies were elites, especially those of science and the state, the one to advise and the other to enforce. It sought to identify proper practices and ensure their use, and it justified both knowledge and power by appeal to positive Reason. The peculiar circumstances of the colonies ideally suited them as an arena for conservation's theory and practice. Europe's elite could do in the colonies what they dared not try at home. They could, in particular, counter the threat of free-burning fire in ways unwarranted and by means unacceptable in Europe proper.

Land Reservation: Bounding the Fire Frontier

To colonize was to shake up land use. There was, however, no predicting how contact might unfold ecologically except that what Europe expected to happen rarely did. But whether the colonizers did the labor themselves or worked obliquely through native peoples or imported slaves, they tried to point resources to their own particular markets, which usually meant "rationalizing" land use and ownership, which, in turn, meant to them that traditional usage had to reconcile with European law, market capitalism, and science. The great divide, of course, was between those lands in which Europeans merely ruled and those they (and their goats, oxen, and dandelions) overran and settled. On which side of the divide they resided largely decided how much land they could remake and how quickly.

But one strategy spanned both conditions. The colonial powers set aside lands, often immense, as public reserves. Typically these were forests. (Originally, "forest" had a legal not a botanical meaning. It meant a reserved area, a place subject to forest law, usually for purposes of royal or aristocratic hunting.) Wholesale reservation allowed colonial officials to at least regulate, if they could not abolish, the waste of commercial logging, errant herding, and folk foraging. In principle, the intellectual authority of science would bond with the political authority of the state to replace both the reckless selfishness of private capital and the communal lethargy of the village. Reserved lands were thus a means not only to conserve but to modernize.

Setting aside lands for the public good worked best on emptied places. A combination of introduced disease, war, and forced relocation often achieved just that: it removed the local peoples and lifted the human hand from the landscape. Where the frontier advanced slowly, where lands held rich soils, where the native population was dense and receded in lockstep with the advancing colonizers, European agriculture reclaimed the sites almost as fast as they became vacant. But elsewhere a great void appeared, as for one reason or another settlement lagged and large fractions of the land were uninhabited.

Precisely at this moment, conservation congealed as a political philosophy. The vacated lands all but begged for state intervention. Reservations for forests, watersheds (catchments), wildlife, national parks and later preserves for scientific research were the outcome. Their dimensions could be significant. A third of the United States is public land, almost two-thirds of Australia is public or crown leasehold land, and a whopping 89 percent of Canada is federal or provincial crown land. Much of Old Russia was "public" because the tsar had claimed new territories during the expansion east; then the Soviet Union nationalized all lands. But these nations are exactly those that dominate wildland fire research for the simple reason that they hold so many wildlands under state governance.

Elsewhere, the reservation policy foundered because people remained in or around the gazetted lands. Perhaps the most celebrated experiment occurred in India, where Britain began "rationalizing" land use for revenue payments and then created large forest reserves in the belief that preserved woodlands would stabilize climate and rivers (hence agriculture) and that only the power of the imperial state could counter greedy marketeers and ignorant villagers. Recognizing that natives needed at least partial access, if only for occasional grazing, colonial foresters established three categories of forest, from full to lesser degrees of protection. The reserves proved immensely unpopular with rural Indians; moreover, even a handful of herders suitably armed with torches could subvert imperial goals. Still, despite the scheme's administrative burden and native hostility, British India boldly pressed on and, decade by decade, shifted more lands into gazetted forests. Proclamations, however, were one thing, practice another. No item more obsessed the reserves' guardians than fire; and no weapon proved more powerful in the hands of those who resisted.

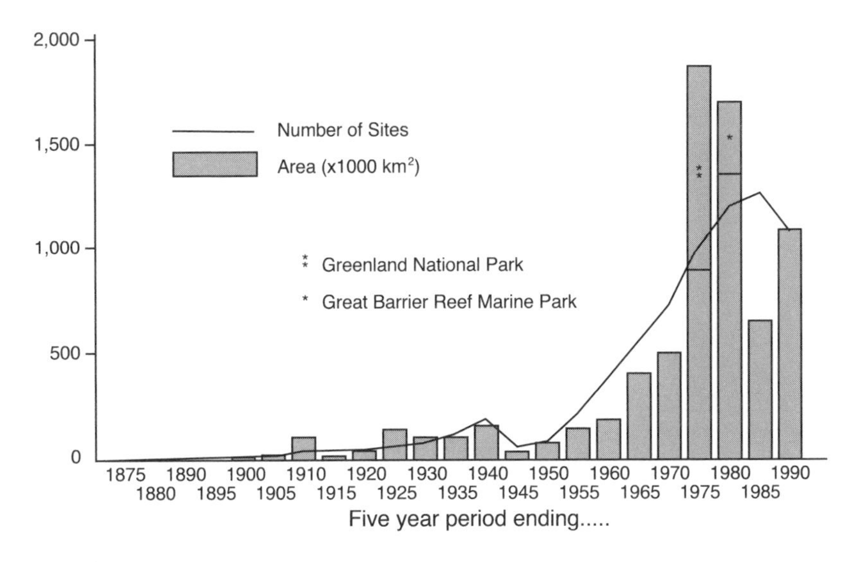

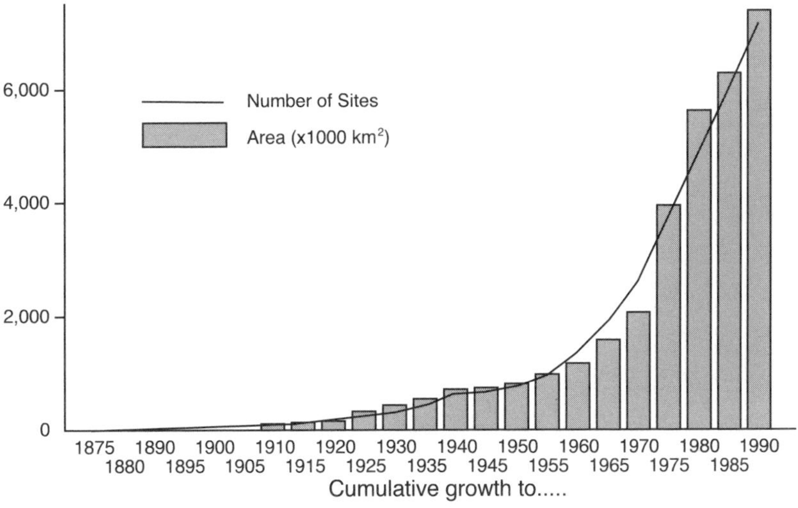

FIGURE 12. The Earth's protected lands. Industrialization has created new categories of land use, perhaps most spectacularly "protected" lands in the form of parks, nature reserves, and public forests. The top graph shows their growth by 5-year increments, the bottom, the cumulative number of sites and area. Not all these lands burn: the great surges of 1975 and 1980 resulted from including Greenland National Park and the Great Barrier Reef National Park, neither prone to fire. But a "protected" classification nearly always means a change in fire regime. While most of the developing world's fires burn in landscapes roughly agricultural or pastoral, most of the developed world's fires occur on protected sites. (Source: World Conservation Monitoring Centre, 1990, redrawn by the University of Wisconsin Cartographic Lab)

The Indian model traveled unevenly. Forest reserves became a common feature of colonial rule, yet underwent various levels of compromise. They worked only by restraining peoples who moved through them with the seasons, or by pushing them out altogether. This proved difficult in densely populated places or demanded more political will than most economy-conscious imperialists were prepared to muster. Such reserves were a conception of the Enlightenment, so did not appear until late (if ever) in Spanish and Portuguese colonies—save Mexico, which acquired the practice by the examples of France, where its forestry officials studied, and of the United States. They emerged in Africa, much compromised, but most spectacularly in the form of game preserves like Kruger and the Serengeti. These resulted from a convergence of drought and famine, a history of slaving and war, and the epizootic rinderpest, which wiped out nearly all cattle, sheep, and goats in eastern and southern Africa and allowed brush, the tsetse fly, and wildlife to restock the vacant landscapes just when European imperialists advanced in force. Elsewhere, forest reserves proved difficult, as galling to locals as national forests proved in the American South. As often as not the friction erupted into fire, so often and so visibly, in fact, that smoke and flame became a public test of whether conservation could do what it claimed and hence deserved the political clout it demanded.

Forestry: The Enlightenment Passes the Torch

Few reserves were set aside except to be used, which meant they needed someone to oversee them, preferably with ties to science, government, water, and trees. That task fell to—or rather was seized by—foresters. Imperial forestry, in particular, glued together three traditions, like veneers into plywood. The Germans excelled as silviculturalists, grafting the cultivation of trees onto the great rootstock of European agriculture. The *dirigiste* French bonded forestry with government, as an institution by which the state could undertake, in the name of the common good, major reclamations on degraded landscapes. Ironically, it was the British, who had no forestry tradition (or, for that matter, forests worth the name), who wrapped these together and shipped them out to their sprawling empire. Forest reserves, forestry bureaus, forest science—the lot coevolved, imperial institutions all, and all haunted by fire.

Forestry thus expanded hand-in-glove with the creation of reserves. Forestry required public forests, while publicly reserved lands demanded wardens, a role foresters claimed as a birthright. So ironically fire, mostly embedded in farms and fields, not temperate woodlands, became the intellectual property, practical domain, and professional obsession of foresters. Forestry became the medium for fire control, for fire science, for national fire strategies. Despite its varied splendor—few things are harder to hold than flame—fire came to be viewed through forestry's peculiar prism.

Everywhere outside cities, foresters became the oracles of fire knowledge and the instruments of fire's control. It was a role they relished, if reluctantly. For all their ritual denunciations of fire and yearning for that future age when flame would vanish, fire engaged them as nothing else could. As a hero is judged by the strength of the villain he struggles against, so forestry grew strong through its contest with flame. Forestry's implacable nemesis was also its glory, its romance, the threat that more than any other granted it political power, the recurring deed that kept it before the eyes of elite and public both. The saying grew, "Scratch a forester and you'll find a firefighter." Had fire not existed, colonial forestry would have been wise to invent it.

How Europe Redefined Fire's Realm

Yet all those reforms in land use remained on the surface. While fire regimes changed, the pieces of landscape mosaics tended to endure, clicking into new patterns with each turn of the political kaleidoscope. The real revolution in fire lay deeper, with the tectonic thrusts of industrialization. The Enlightenment had justified replacing folk practices with scientific technology, but only outright industrialization furnished the means to make it truly happen, while Europe's global imperium provided the medium to carry it around the Earth. In varying rates, at divers times and places, the new pyrotechnics began to add to and then shove aside traditional fire practices. Officials and the scholarly classes actively campaigned for the exchange. The world began to fission into two great spheres, one that burned biomass, the other that combusted fossil fuel. Only in select places, and then perhaps only temporarily, did the two coexist.

Industrial Wildlands

Perhaps the most interesting transition occurred in the reserved wildlands. Domestic fire, transporting fire, agricultural fire—these responded to economic pressures and, for the household, to concerns over public safety. Closed combustion (or electricity) was usually cheaper and healthier than open flame or smoldering coals. But it was not obvious what administrators should do with crown forests, national parks, wildlife reserves, and the like. With people no longer living off the land, the habitual sources and reasons for fire had vanished. While a certain number of transients set fires, travelers, poachers, and tourists set nothing like the number that had previously abounded, nor did these fires occur along the lines of the old matrix. In most colonies, too, lightning defiantly continued to kindle fires without regard to legal bans on burning. In brief, fires continued.

The almost universal response was to attempt control over all flame, to suppress its starts quickly, and to banish it wherever imaginable. Third Fire industrialization seemed to promise that such a scheme was possible. Motor vehicles, aircraft, portable pumps, rubber hoses, and chemical retardants—all allowed firefighters to rush to fire outbreaks and apply a powerful check. It was a simple matter to beat down flames if they were detected soon enough, so officials erected lookout posts and laid down roads. For larger fires they mounted military-like campaigns to surround free-burning fires, choke off their supplies of fuel, and mop up every ember. So here, too, industrial fire substituted for open fire, though more slowly. Even where bureaus pursued the strategy zealously, as in the United States, the system took decades to mature to the point where burned areas plummeted. But, smoke by smoke, it did.

What changed was not only fire's regime but its sheer presence. Fires set by residents dwindled. Fires kindled by lightning or accident were suppressed quickly or fought off hand-to-hand for days. Fires that formerly lingered for weeks, maybe months during dry seasons, creeping and sweeping with each puff of wind, were snuffed out. Symbolically, fire even ebbed as a technique of firefighting. Previously, large fires were attacked by backfiring from a river or road or ridgeline. When backfiring worked (not always), it checked the spread of wildfire. Yet win or lose, the practice kept free-burning fire on the landscape. More and more, however, agencies attacked fires directly, matching the wild force of the flaming front with the contained counterforce of industrial combustion. Fire in all its forms began drying up like a desert pond.

Two Worlds of Fire

Still, it was not possible to remove fire without consequences. Burning continued wherever the underlying conditions permitted, and its expulsion (or attempted expulsion) caused often unexpected and unwanted side-effects. Stubbornly, temperate Europe continued to declare itself as a norm, and because of its global reach—its imperialism, its hold over modern science, its industrialism—it confirmed its own odd standards as those best suited for the planet. Fire remained suspect, open burning survived as a stigma of primitivism, and the abolition of flame endured as an ideal of land stewardship.

The present geography of fire thus shows a striking imbalance. Most open fires occur outside the sphere of Europe's influence. With a few exceptions, the amount of flame is almost a measure of political or ecological resistance to European colonization. (Boreal Canada and tropical Australia, for example, continue to burn, but outside the realm of European agriculture and population clusters.) The primary body of fire expertise today resides with those former colonies that combine science with wildlands. However strangely, given Europe's history, it continues to control the flow of organized exchanges between nations, as for instance the UN's Food and Agriculture Organization's (FAO) technical assistance program and fire study tours. Those nations that don't burn are telling those that do how to stop.

The European frontiers of fire thus abide. Because fire ecology is about ideas, information, and institutions as much as it is about fuel loads and seed banks, the European hegemony (broadly and historically interpreted) is not likely to end soon. Europe's expansion launched a Great Transformation in the fire history of the Earth, the most significant since the retreat of the Pleistocene ice. That impact persists, like lands still rebounding upward from the ice's release. Long after Europe's empires have shed their substance, their shadows still cover most of the Earth.

If the focus of fire protection has shifted from the European core to the neo-European pale, if the neo-Europes in North America and Australia are keen to restore some fire to their wildlands, Europe nonetheless continues to influence global fire management by its wealth, its industry, its science, its canon of environmental values. Even Japan, never colonized but modern, accepts fire practices more or less indistinguishable from those of temperate Europe's. The Kyoto Protocol, which seeks to regulate the production of greenhouse gases, did not originate with nations rich in flaming savannas and smoking swiddens but with those belching coal and choked with the exhaust of automobiles.

Yet the desire to modernize will surely propel the fire-rich nations to drop their torches, as Europe did, in the belief that the two fires cannot

Chapter Nine

Industrial Fire

STOKING THE BIG BURN

Getting spark and tinder together at the right moment was, for nature, always chancy. Humans improved those odds by making ignition more or less constant and by chipping or coaxing biomass into ready fuel. This did not ensure, however, that lands burned at will. Not every spark kindled flame; not all combustibles could burn at all times; weather mattered enormously. Outside of dwellings, fires still burned with the seasons. But the primary limitation on humanity's control over fire remained having enough of the right stuff to burn.

No matter how clever a people or how ingenious their technology, fire could flourish only where biomass could. Anthropogenic fire could not evade the ecology of growing plants, could not escape the cycles of life and decay, or could do so only for a while and then with serious damages. People could fashion grown biomass into fuel, but they could not make fuel from nothing. They needed new worlds for fire. Europeans did that first by geographic discovery, then by a technological one, the discovery of ancient lands, long fossilized but now ripe for burning. The outcome was industrial fire.

What does "industrialization" mean? Commonly it is understood as a social, economic, and perhaps political process that redefines the relationship of people to one another. Only secondarily has it been considered an environmental event, and then murkily, as a source of pollution. But its meaning for fire history is crystal clear: it refers to the burning of fossil biomass. Just as Second Fire had before it, industrial fire sought out or created new landscapes for burning, and so expanded fire's realm. Humans could now burn biomass stockpiled over geologic time, a millionfold increase in fuels available for combustion.

The source mattered because even prime movers like steam engines and their offspring could not by themselves shatter the primordial ecology of fuel so long as they burned wood, peat, or dung. They remained very much within those old cycles and, being ravenous, only rushed the fuel question to a crisis. The technology mattered, too, because coal and oil demanded a suitable combustion context. Unlike branches, sod, and seaweed dragged to fields to overlay fallow, they contributed nothing if burned in the open. Third Fire was no less interactive than those that preceded it, but unlike them it burned within a technological setting rather than a natural one. Combustion, fuel, and machinery thus co-evolved, as flame, vegetation, and air had before.

If what went into the flames differed, so did what came out of them. Industrial fire began to alter every fire habitat, overloaded ecological sinks, and reshaped the society that wielded it. It affected not only fields, farms, woods, and wildlands, but cities, manufacturing, trade, capitalism, politics, technology, and social order—all on a planetary scale. What "industry" meant after Third Fire was very different from what it had meant before. So was our concept of the human role as fire keeper.

In older fire ecologies, everything humans did could be done by something else. Lightning set fires, elephants pushed over trees, wombats dug in the ground, bison fed on grass, cougars hunted deer. Humans had an extraordinary capacity to mold and move the pieces of this mosaic, but if they left, those parts would assume, by themselves, some new pattern. In a fundamental way humans could depart the scene and the basic principles of fire ecology would still apply. This is not true of industrial fire.

FIGURE 13. The Big Burn. The defining trait of Third Fire is its reliance on fossil fuel. That required, in turn, new chambers to combust the mineral biomass, a combination that broke down and isolated fire into its elemental features. For Earth, Third Fire announced another defining trait, that this species of combustion depended utterly on humans. The top graph tracks the outcome, the Big Burn, as measured by the annual flux of carbon from burning fossil biomass. In fact, the modern world's reliance on fossil biomass is greater than these figures indicate because petroleum, in particular, is distilled into other chemicals for uses other than fuel. By any reckoning, the burning of fossil biomass constitutes a new source of earthly fire on a huge and escalating scale. Calculations from 1990 estimate that Third Fire claims some 60 percent of the planet's overall combustion budget. How this combustion interacts with the other sources is complex. Mostly it has enhanced First Fire (by stimulating nature reserves) and has demoted Second Fire (by offering other fire technologies and actively attacking free-burning flame). The bottom graph shows the relative power of living and fossil biomass as sources of power. Absorbing this immense combustion load has profoundly upset the planet's ecological networks. (Sources: King et al. 1991 and Smil 1994, both redrawn by the University of Wisconsin Cartographic Lab)

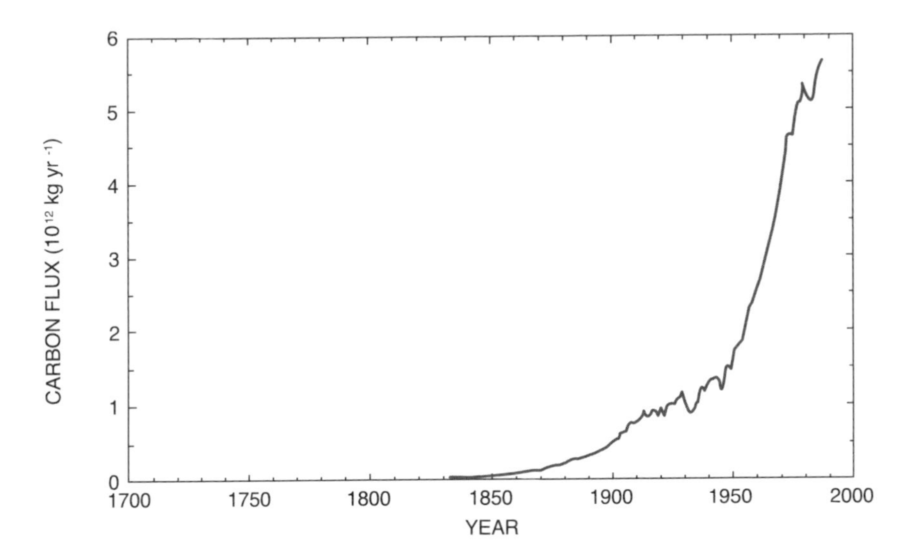
CARBON FLUX (10^{12} kg yr^{-1})
YEAR
0
1
2
3
4
5
6
1700
1750
1800
1850
1900
1950
2000

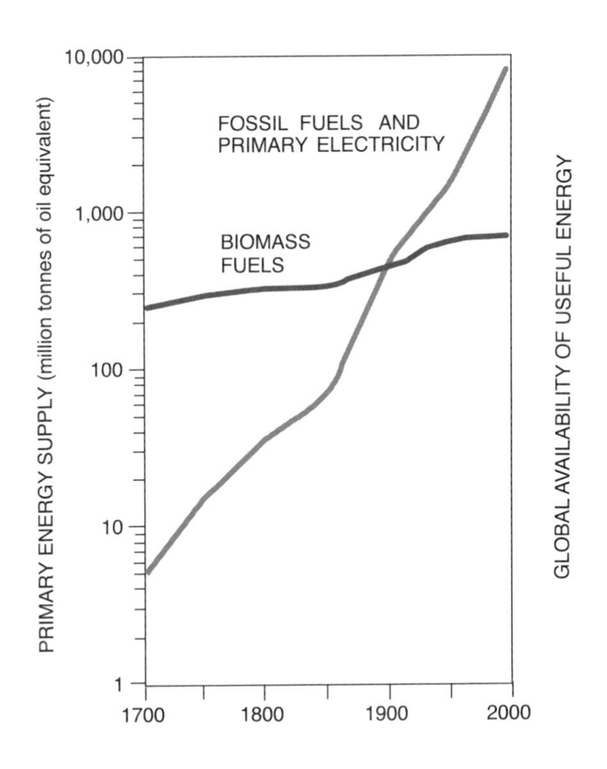
PRIMARY ENERGY SUPPLY (million tonnes of oil equivalent)
GLOBAL AVAILABILITY OF USEFUL ENERGY
FOSSIL FUELS AND
PRIMARY ELECTRICITY
BIOMASS
FUELS
10,000
1,000
100
10
1
1700
1800
1900
2000

It can occur only with humans as agents. The act of exhuming, burning, distilling, or otherwise processing fossil biomass is humanity's alone. If people leave the scene, the principles of industrial fire leave with them. It is possible to imagine an ecology of free-burning fire without humans; it is not possible to imagine an ecology of industrial fire. Humans are not simply disturbers: we are what make the system work. The power and the glory of Third Fire—along with its flaws and disasters—are ours alone.

How Industrial Combustion Has Added Fire

All this changes what is burned and how combustion, its fuels and byproducts, pass through ecosystems. It has always been that more fuel means more fire, and so it is with Third Fire. Industrial combustion has added enormously to the Earth's fire load; probably there is more combustion on the planet than ever before.

The Era of the Big Burn

In fact, this huge transfer of fossil biomass involves more than simple burning. Part of fossil biomass ends up as fossil fuels, fed into prime movers, but other parts become fossil fallow and fossil forage, feeding both fields and machines. Whatever remakes fossil biomass also transfers carbon and other substances from the geologic past into the present, and since the present is often unable to absorb it, the outflow spills into the future. Outright burning is only one means; but nearly all the others involve combustion in some form at some stage, if only to break down, distill, or otherwise convert buried geochemicals into active biochemicals.

The exhumed fuels, being burned, add directly to the Earth's overall fire load, while the exhumed fallow and forage contribute indirectly by allowing us to "burn" farms offsite and to plow fire's products without its flame. This input shatters (or to state it more aptly, transcends) the old ecological cycles: putting more in means that we can take more out. Agriculture can intensify in ways not possible before. In place of the fumigating and fertilizing fire, we can apply artificial pesticides, herbicides, and composts, ultimately derived from fossil biomass. The critical combustion occurs in furnaces and kilns rather than with flame spreading across fallowed fields and rough pasture. The fossil forage of gasoline and diesel can feed a mechanical menagerie of automobiles, trucks, tractors, motorcycles, pumps, generators, ships, and aircraft, all of which (and more) make up the food chain, and the respiration cycle, of industrial ecology.

So profound and so extensive has Third Fire become that it has replumbed the flow of combustion throughout the planet. The fire regimes of the Earth are increasingly those of industrial combustion, or of places wrenched or welded by it into mongrel landscapes. Fire now passes over oceans, through the skies, even to other planets. The amount of fossil biomass burned currently exceeds that of living biomass. Its proportion is likely to rise in the future: there is a huge reservoir of fuel to tap and most of the world's population has not yet begun to burn it in bulk. For probably another century or two, the Big Burn will almost surely rule the combustion regimes of the Earth.

The Big Dump

But if the sources are new, the sinks are not. The Big Burn is creating an equally huge Big Dump. There is no geologic midden for the byproducts of industrial combustion that is comparable in scale to the coal beds and oil fields from which the fuels come, none at least that can soak up wastes as quickly as Third Fire spews them out. The smoke, ash, and gases of industial combustion must lodge in the same places as the output of other burning. These have rapidly overflowed, and in a short time—a geologic heartbeat—the Earth has found itself awash in an extraordinary spillage of pollution.

Equally to the point, the ecological links that long joined source with sink are breaking down. First and Second Fires burned within a biotic world that promoted and absorbed fire. Third Fire does not. Industrial combustion burns without regard to cycle or season or fire-tempered adaptability. It puts fire where it had never existed and removes it from places that have long known it. Perhaps more provocatively, the tremor of the Big Burn may herald a series of combustion aftershocks. Greenhouse gases gushing out of industrial burning threaten the stability of the global atmosphere, which could easily redraw the geography of surface burning. Less apparently, that ancient carbon broken out of its bondage by burning is not inert; it interacts with the living world. An airshed enriched with carbon dioxide will spur plant growth and may plump landscapes with larger stocks of peat and woody biomass.

As in the past, flame will likely follow fuel. Free-burning fires may become more intense, the costs of containing them higher—not least the cost in more carbon output through the further combustion of fossil fuels. The burning of ancient biomass is thus unfettering carbon that may become fixed into living biomass that can, in turn, burn and reburn. The Big Burn will not ban fire. It may, at best, reshuffle it and at worst redouble its presence.

Big Ideas

As important as the flow of combustion is the flow of knowledge. Exploiting fossil biomass has shocked ecological models as thoroughly as discovery of the New World did the cosmological models of medieval Europe. The frontiers of ecology's old world of knowledge are dissolving. Against the command to recycle the received wisdom of the Ancients stands the possibility of more and better knowledge. An agricultural model has long underpinned ecological theory; the garden has served as synecdoche for the globe. Knowledge, like nutrients, seemingly cycled and recycled endlessly.

But industry has so far proved additive, progressive rather than cyclic. Technology grows, knowledge grows, biomass consumption grows. All that inherited ecological theory can do is condemn it, declare its growth unsustainable, track its wreckage of existing biotas, and speak with expectant irony. The carbon wealth plundered from the past, it argues, will push nature's economy into an inflationary spiral as surely as the sacked bullion of the Aztecs and Incas did 16th-century Spain's. Or it may be that the Earth can absorb more than theory allots, and that wholesale tinkering with the biosphere and atmosphere may in the end be necessary. What the Big Burn means and how it might continue is vague. It may be that its fossil fires will be banked, or that like a flaming front, shallow but intense, it will pass over the Earth and leave a new growth of knowledge to flourish amid its ashes. This much seems clear: the challenge posed by industrial fire to biological theories is no less than the impact fire practices are having on real landscapes.

How Industrial Combustion Has Subtracted Fire

Third Fire has done more than add its hidden flames to the Earth's combustion stew: it has also eliminated fires other than its own. A house, for example, does not need two heating appliances, one a wood-burning hearth and the other a gas-fired furnace. The cheaper, safer, and more efficient devices of industrial combustion are triumphing over open flame through a process of technological attrition. But in part substitution has also involved active annulment. Third Fire pyrotechnologies have systematically suppressed their combustion competitors.

Transmuting Fire: How Third Fire Substitutes for Second

In realm after realm of human technology, industrial fire has shoved out its flaming forebears. Mines no longer roast ore and smelt metals over open flame. Chemists no longer simmer, distill, and boil their broths over fires. Builders have ceased to dry timber, bake tile, roast limestone, and fire bricks in ovens stoked with wood or charcoal. Special furnaces now burn gas or, better, heat with electricity generated by combustion-powered plants far removed from the factory. Where almost all technology had once been overtly a pyrotechnology, now little is. It appears—if fire appears at all—as a more rarefied combustion. That transformation has rippled everywhere throughout modern societies.

Fire has all but vanished from houses. An American home is more likely to feature an electronic entertainment center than a functional fireplace. The family-gathering hearth has dissolved into the virtual village of television. Lightbulbs have replaced candles and whale-oil lamps; gas or electric ranges, the wood-burning stove; the microwave, the teapot whistling over a flame; flashlights, ember-dripping torches; central furnaces or electrical space heaters, the central fireplace around which a family would huddle. Even candles have shrunk to the realm of ceremonial birthday cakes. It is possible to live years in a modern house without ever seeing the fires that once, almost by definition, made a house a home.

Outdoors, the same pattern prevails. More and more, homeowners are unlikely to burn leaves or pruned branches. Careless fires might escape; smoke can annoy neighbors, who themselves burn little. Exurban migrants are repopulating rural landscapes, and planting urban values and expectations along with their daffodils. Fire codes regulate open burning to the point of outright bans. City and suburban residents prefer to run power rakes over dead grass and then bag the debris in plastic sacks to haul to landfills, which bury rather than burn the refuse. Air quality considerations have eliminated celebratory bonfires; they threaten to restrict even charcoal barbecues. Urban residents—and most citizens of industrial countries live in or around cities—can pass years without seeing a fire except as a disaster or an image on a TV screen. The vestal fire is now little more than a virtual fire.

Urban landscapes have actively sought to expunge flame. Modern building codes, noncombustible materials (often manufactured with industrial pyrotechnics), cities platted with wide streets for automobiles, vigorous firefighting institutions—all have dampened the presence of fire. It appears mostly by accident or arson, and almost always as a danger. Paradoxically, fire's influence endures, more vigorous than ever. Virtually every niche of the built landscape—every room, every structure, every city block—follows designs intended to prevent fire, or if a fire begins, to stymie its spread and provide easy escape for residents and access for firefighters.

But the process has not ceased at the city limits. Steadily, industrial fire has shorn away the once-essential practices of agricultural burning. The biotic jolt a hot fire had previously given a fallowed field, a flaming front of chemicals now does in the form of pesticides, herbicides, and artificial fertilizers. The flames that once churned the land now crank the wheels of diesel-powered tractors. So farmers cease to fallow, drafting that burnable biomass out of the geologic past, and as those scruffy fields fade so does their biodiverse legacy. The once-routine burning of cuttings, garbage, irrigation ditches, and pastures has become infrequent, even quaint, and vaguely disreputable. If the fields or ranches reside near cities, urbanites complain about the smoke from open fires as both a

---

FIGURE 14. Missing fires. The industrial countries have far less fire now than they had in the past. The United States shows why. One cause is that officials actively suppressed fires of any origin on public lands and nature preserves.

The top graph shows the outcome for America's wildlands. It tracks three categories of land—those protected by federal agencies, those by institutions that cooperate with the federal government (mostly the states), and those lands as yet unprotected. It becomes immediately obvious that most burning occurred on the unprotected lands. Installing a first-order apparatus to control fire under these circumstances can be most successful—at least for a while. Even as land is transferred from unprotected to protected status, the total acreage burned plummets. The economic and ecological costs, however, have been significant.

The bottom graph shows the complementary cause for fire's ebb: the recession in active burning, here by ranchers in California brushlands. The figures make a critical point, that it was not simply that natural or accidental or malicious fires were being extinguished but that controlled fires were no longer being set. The erasure of free-burning fire from the American landscape resulted from both causes. (Sources: Bureau of the Census 1975 and Biswell 1989, both redrawn by the University of Wisconsin Cartographic Lab)

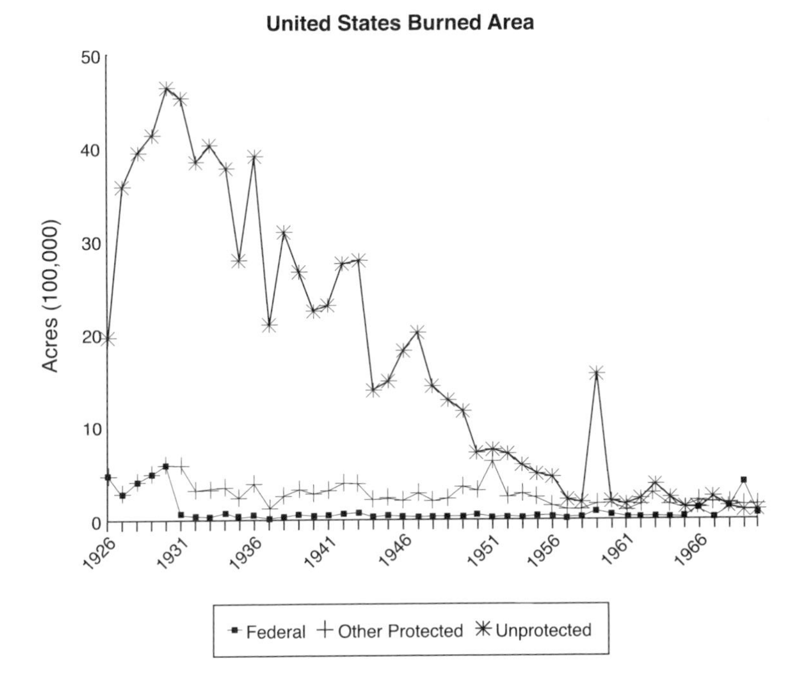
United States Burned Area
Acres (100,000)
50
40
30
20
10
0
1926
1931
1936
1941
1946
1951
1956
1961
1966
Federal
Other Protected
Unprotected

California Burning By Permit

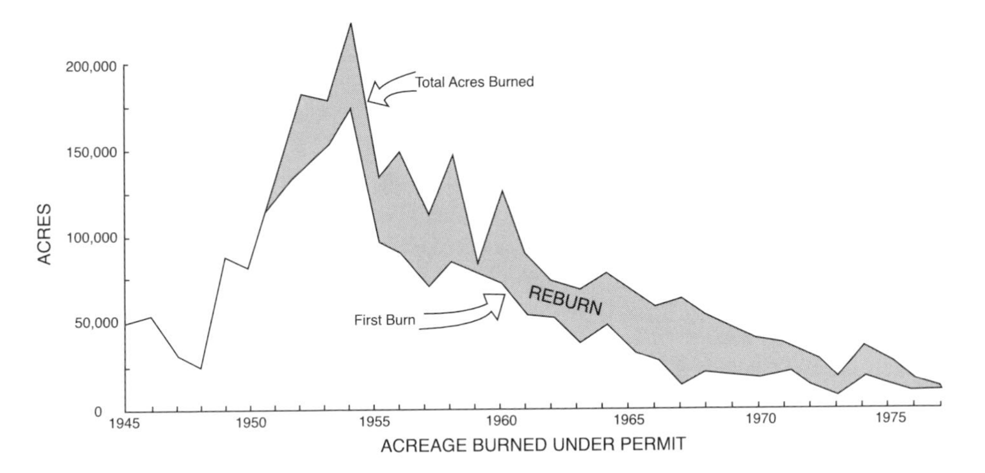
200,000
150,000
100,000
50,000
0
ACRES
Total Acres Burned
REBURN
First Burn
1945
1950
1955
1960
1965
1970
1975
ACREAGE BURNED UNDER PERMIT

nuisance and a threat to public health and condemn the burning as a primitive relic, a rural superstition, or environmental vandalism. Burning wheat stubble, grass seed plots, and rice residue all now suffer bans, more or less complete.

The record suggests, however, that rural fire habits are well rooted and that removing them is tricky. Fire-fallow agriculture relies on flame as a more robust technology than concentrated heat. Replacing is more formidable than the model of a stove implies. Third Fire farming does not unseat Second Fire farming as simply as gas replaces wood in a furnace. Nor does it respond to direct assault by official prohibitions, condemnations, or even firefighting. Burning is not a moral choice, one option among many, a measure of sloth or ignorance, but an ecological mandate without which the fields decay. Fire-fallow farming endures as long as its population of farmers does, and they remain until they are slowly siphoned off to cities by the pull of industrial jobs or forced off the land by eviction, enclosure, plague, or collectivization. In the industrial transformation, farming follows rather than leads. The full cycle may easily take 50 to 60 years.

While all this has generally improved public health—smokey rooms, for example, invited respiratory ills—the upshot is that few urban residents have firsthand experience of fire or know it outside the built landscape. They understand it as a technology for which other, more advanced technologies can substitute. They understand it as a danger and a hazard that proper codes and materials can help contain. They

---

FIGURE 15. The two geographies of earthly fire. Today the world is dividing into two broad realms of combustion, one fueled by living biomass (top map), the other by fossil biomass (bottom). The latter coincide almost exactly with industrialization. Note the difference between Europe and Africa (graph). A few places exist that show neither (such as the Sahara) or both (India, Indonesia). Those with both exist because, while Third Fire is booming, village life continues over substantial swaths of countryside, and the village economy remains rooted in biomass for fuel, farming, and herding. In general, Second and Third Fire do not coexist willingly. Paradoxically, Third Fire, by promoting nature reserves, has promoted First Fire. The likely future is that Third Fire will continue to drive off open burning of all sorts. This comes with ecological costs: not only the need to find sinks for combustion effluent as vast as the sources that power industrial combustion but the costs of removing fire from biotas long adapted to them. (Source: Lim and Renberg 1997, redrawn by the University of Wisconsin Cartographic Lab)

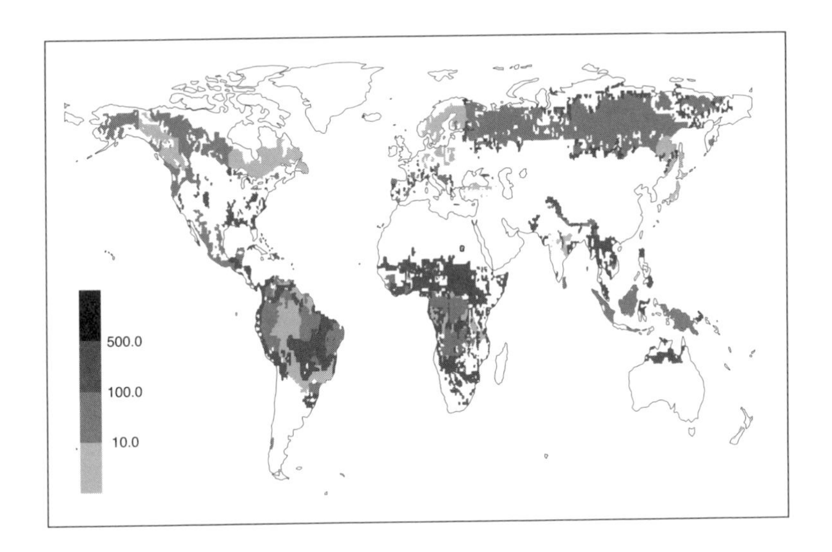
500.0
100.0
10.0

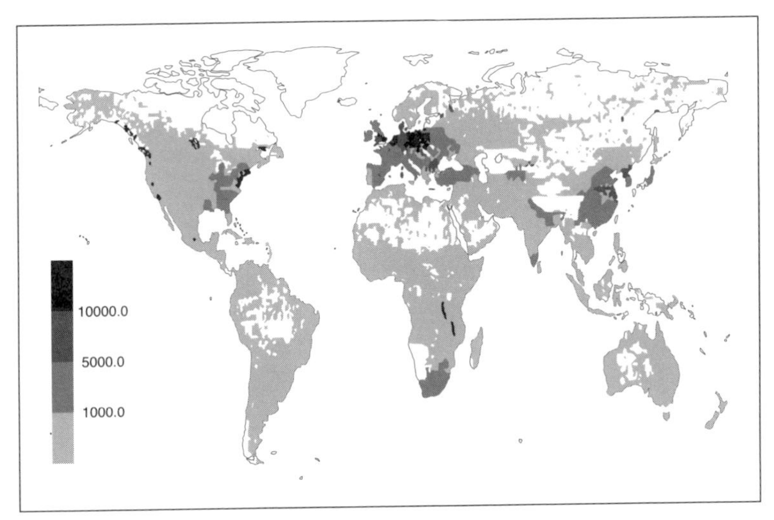
10000.0
5000.0
1000.0

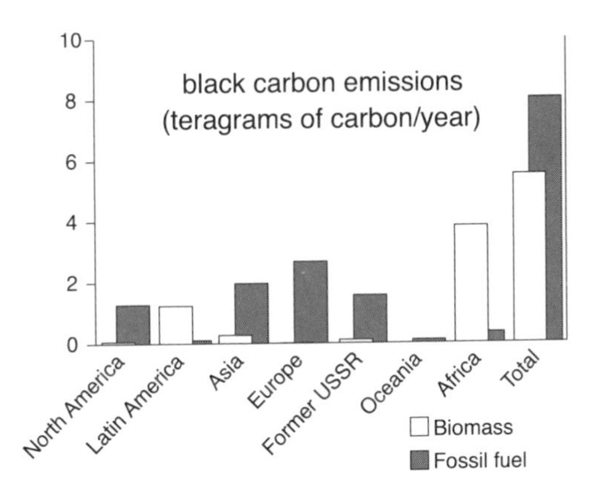
black carbon emissions
(teragrams of carbon/year)
10
8
6
4
2
0
North America
Latin America
Asia
Europe
Former USSR
Oceania
Africa
Total
Biomass
Fossil fuel

experience it three or four times removed from its source—through electrical appliances, through LCD screens, through their automotive "ignition" keys, and through pollutants that darken the sky and clog the lungs. That fire might be more than a tool or a nonmechanical technology is something they appreciate, at most, with their minds, not their hands. Humanity as keeper of the flame has become a shopworn cliché, not a metaphor that strikes to the core of our ecological being. If fire is a device, they want an improved, flame- and smoke-free upgrade. If fire is somehow ecologically essential, they wish to confine it to a suitably remote ecosystem.

Banned Burns: How Third Fire Suppresses Second and First Fires

The Grand Exchange of Third Fire for the others extends also to the attempt to strike fire from wildlands. With less land committed to fallow (and other lands not suitable for farming in any event), industrial societies have been able to endow new landscapes, notably nature reserves and parks. Initially, it seemed only proper to protect them from wildfire, particularly from fires set by vagrants and arsonists. After all, this was a nature reserve, where the human hand, including the torch it held, should be stayed or at least hidden. But officials went further to argue for the supression of *all* fires, which they condemned as intrinsically damaging, unsightly, and unnecessary. All-out fire control on wildlands has become a distinguishing mark of industrial societies.

Once begun, the endeavor can rush forward with remarkable power. The usual strategy includes removing native peoples, farmers, and herders, the source of most fires; the erection of a fire-protection infrastructure by which to detect and rapidly attack fresh fire outbreaks; and the systematic suppression of any and all flames. Perhaps not surprisingly, effective control relies on Third Fire machinery. Spotting fires, moving firefighting forces, mustering pumps, bulldozers, and aircraft—all depend on internal-combustion engines, and all oppose the power of free-burning fire with a counterforce of industrial combustion capable of strangling fire from the landscape entirely. Even protective burning and backfires disappear as means of control. Open flame of all kinds dwindles.

But here the fire ecology of the city splits from that of the country. In the built environment, controlled fire is a technology and open fire a hazard. What flame does, some other device or process might do as well or with fewer side-effects. In natural landscapes, however, fire is also an ecological process with a value quite apart from what humans grant it. What it does nothing else can do as fully. Removing a fireplace in a house might be inconvenient but it does not cause the house to decay. Removing fire from many ecosystems does. Replacing a hearth with an electrical space heater might lose something of fire's poetic reverie, but the heater may well be more efficient and safer. Replacing free-burning fires with bulldozers, chain saws, herbicides, and nitrogenous fertilizers, however, can exact serious ecological costs.

And that is what has happened wherever industrial combustion has attempted, either with finesse or by brute force, to replace biomass-driven burning. If adding Third Fire has its costs, so does subtracting First and Second Fires. Not always, not often obviously, yet eventually the extinction of open fire causes biotas to adapt, and compared to what preceded fire's removal, the landscape may come unhinged. Even so, industrial fire practices do not, ultimately, remove fire. Although brief, the historic record is clear that the sudden arrival of Third Fire can stun a biota and that for a period of time—a few years, a few decades—the amount of area open-burned plummets. In many places, however, a restless, ever-tinkering nature, as tireless as the tides, moves on, growing and rearranging itself in ways that promise to bring fire back in one form or another. Nature is not something people have made. It has its own fire logic, separate from ours. It accepts neither substitution nor suppression.

Humanity's deliberate choices then are two. We can either convert those fuels into less combustible forms or begin a program of controlled burning. The default option is to suffer bouts of wildfire and to watch the landscape slide and lurch into something very different from that which a program of protection set out to preserve. Even when absent, fire declares its transmutational powers. The fire frontier between Third and Second Fires is as rough and uncertain as that between Second and First Fires before it.

How Industrial Combustion Has Rearranged Fire Regimes

Few places have escaped Third Fire. Even the Pacific's abyssal plains and Antarctica's ice sheets hold its soot and trap its free-floating carbon. By the end of the 20th century, Earthly geography showed a great partition of fire ecologies. On one side burned living biomass, either from natural or human causes. The other burned fossil biomass. Only in a handful of sites do the two truly coexist.

Where Industrial Fire Rules

The region that created industrialization is also the place where it has worked its fire logic most thoroughly. Traditional burning has virtually vanished. Second-millennial Europe itself resembled a spreading fire, burned out in the center, active along its perimeter. Some 90 percent of the area burned is concentrated along the Mediterranean rim. Yet even here, an industrial conversion is under way. Elsewhere the great smoke palls from the seasonal burning of fallow and peat that once smothered European cities now descend from the poor industrial combustion of sulfur-laden coal and lignite. The transformation is almost total.

The strength of industrialization is one reason, of course. Industry developed in Europe early, found abundant deposits of coal, and lodged easily within European institutions. But the long-standing character of European fire is also significant. Temperate Europe has no pronounced fire season. Except along the Mediterranean and, infrequently, its drought-blasted boreal and continental fringes, Europe has lacked the wet and dry pulsing that traditionally underwrites the natural geography of fire. Fire has existed because people have put it there. It is less a part of nature, like wind and sun, than a handy tool, a servant like plow horses or milch cows for its human masters.

In such circumstances the industrial conversion can be relatively complete. Evening satellite photos reveal a Europe ablaze with electrical lights, not flames. Relentlessly, industrial combustion has displaced open fire in nearly every technology and habitat, including the agricultural countryside where it had most resided. It survived along the Mediterranean because the climate favored it, because it had been so long a fundamental part of those biotas, and because social and political factors kept traditional agriculture alive. Elsewhere, combustion competition is weak. There is little opportunity for natural fire to reassert itself. Remove industrial fire, and anthropogenic fire will return only so long as people choose to live off that land. Remove anthropogenic fire, and natural fire would creep back only in selective niches. A postindustrial Europe has become what it had been prior to the Neolithic revolution: an anomalously fire-free patch of Earth.

Yet that exception matters. As Europe expanded, it assumed, then asserted, its own fire geography as normative. To its officials and intellectuals, how fire behaved in Europe was a standard for how it should behave everywhere. The continuing conversion of Second Fire to Third provides a further gauge of European influence around the world and would seem to vindicate Europe's self-declared standing as a fire authority. So it has come with special force that fire's attempted ban has caused an unexpected crisis in European ecology. Without their catalytic fires, traditional landscapes have begun to unravel and their venerable biodiversity to slump. These were never "natural" landscapes. They were places made with fire-abetted farming and herding. To preserve their plants, animals, and scene would require the preservation of their formative practices. A city block could successfully exchange pine knot torches and oil lamps for fluorescent lighting. The countryside could not.

Where Biomass Burning Endures

There remain places that have so far been spared industrial fire, or have known it only obliquely. The conditions that made Europe's conversion rapid do not apply. These are landscapes with powerful wet-dry cycles and long histories of anthropogenic burning, places that resisted European colonization either demographically or agriculturally, that may lack rich deposits of fossil fuels. Large slabs of Latin America, long swaths of boreal forests in North America and Eurasia, the tropical savannas of northern Australia, much of southeast Asia, and most of Madagascar continue to burn—a significant chunk of the Earth and for particular regions a ruling fraction of their landscapes. Some places burn because people have little control over fuels or spark. The boreal forest, for example, burns because agriculture does not work, logging only adds to fuels, and lightning fires are very expensive to suppress. Other places burn because humans require it, mostly for agriculture. This is the case with the Earth's most impressive display of free-burning fire, that which occurs in sub-Saharan Africa.

Fires explode over Africa like stars sweeping across a nebula. Environmental conditions are ideal. Seasonal rains grow fuels, and seasonal dryness readies them for flame. Still, not every place burns every year. There are deserts like Namibia's and Eritrea's too dry, rainforests like Liberia's and central Congo's too wet. These burn only when rare rains or droughts jolt the biota and unstick the wheels of a normal fire cycle. There are regions like Kenya and especially South Africa that European settlement and industrial economies have penetrated sufficiently to squeeze out or drive off routine firing. While biomass continues to burn, its relative ecological importance shrinks. And cultural factors often prevail. With 125 million humans and major petroleum fields, Nigeria would seem a candidate for rapid industrialization. Yet the oil goes elsewhere and 80 percent of its landscapes—everything between the Sahara fringe and the coastal mangroves—habitually burns.

In Europe, industrialization often replaced older technologies much as silica replaces lignin in petrified wood. Ancient fallow substituted for modern; the previous grain of the landscape remained. In the Neo-Europes like North America and Australia, industrialization joined the grand swarm of peoples, plants, animals, and institutions that had swept over and redefined continents. But this did not happen in Africa. Agriculture in Africa did not have a ready industrial substitute. Colonization through immigration could happen only selectively. Industrialization demanded more than a transfer of tools, ideas, and fire appliances: it required a reformation in the way of life. That has come slowly and spottily. Instead, industrial fire huddles into cities or watches its fuels be exported. The landscape continues to flame openly, as it has for tens of millennia.

Where Biomass Burning and Industrial Combustion Both Thrive

One could reckon, based on the historical record, that Third Fire will eventually drive out the others, that the immense stocks of fossil biomass will overwhelm other fuels, and the flood-tide flow of industrial combustion will overlay the Earth's energy pathways. Clearly this has happened across many lands, and the process of conversion will continue wherever conditions permit. But that may not occur everywhere.

There are places where biomass burning and industrial combustion coexist. It is not clear that this mixture is unstable, or that it is only a notch toward inevitable Third Fire standing. In Mexico, India, and Indonesia, for example, both industrial and anthropogenic fire may abide side by side for a considerable time. The critical factors are a suitable environment for burning, a large rural population committed to fire-fallow agriculture, spotty (if dense) cities that thrive under industrial fire, and ample reserves of fossil fuel. The presence of enduring populations that live off the land as farmers and herders—particularly if granted special favors or legal standing by the government—ensures that anthropogenic fire will remain until those peoples find some other livelihood. Where their numbers are large, the Great Exchange may not occur any time soon.

But a second kind of mutual presence has emerged that joins Third

Fire with First. It arises because industrialization can unshackle substantial portions of land for uses other than cultivation. A goodly number of places, such as nature reserves, have plentiful fuels and ample lightning, and can burn briskly on their own. They *expect* fire, and may suffer and unravel in its absence. No fire appliance can do for them what free-burning flame does. Recognizing this fact, some developed nations have sought to oblige natural fire within their management regimes. They allow naturally ignited fires to burn with little more control than light-handed oversight. Probably, though, this mix will prove unstable. Officials may find it necessary to do most of the required burning themselves. The eventual outcome could well resemble a tossed salad of fire practices built out of controlled burning and hardcore suppression, and seasoned with a sprinkling of natural fire.

The legacy of Earth's ancient ecology will thus check—or at least give pause to—the prospect of Third Fire's unseemly sprint to combustion hegemony. Like anthropogenic fire before it, industrial fire for all its power rests on conditions over which it may triumph but cannot transcend. It cannot substitute for all of free-burning fire nor fully control the remainder. Increasingly industrial societies are deciding it is not worth their while to even try. So although their relative proportions will vary over time and across geographic regions, the three fires most likely will endure, each supreme in its distinctive realm, each quarreling along shared borders, all overlapping in awkward and unexpected ways.

Chapter Ten

The Future of Fire

BURNING BEYOND THE MILLENNIUM

To casual observers, 1997–98 was, as the World Wildlife Fund declared, "The Year the World Caught Fire." Flames seemed to erupt everywhere, and what didn't burn outright appeared to vanish in a planetary pall of smoke. A climatic shift almost tectonic in power, the most extreme weather in a century of records, reversed the normal flow of the Pacific Ocean's El Niño–Southern Oscillation (ENSO). Normally humid areas dried, and arid sites wetted, both creating fuel. Where humanity failed to supply the spark, lightning succeeded.

The scale was breathtaking: the Pacific became a true ring of fire. Some 2.5 million hectares burned in Russia's Far East, almost 5 percent of the Khabarovsk region's forested estate along with most of the northern half of Sakhalin Island. Crown fires broke out in December in the normally snowed-in forests of Alberta, then raced through an early spring. Another immense swath of fires burned in Indonesia, from Sumatra to Java to East Kalimantan. Wildfires broke out in Australia. Rainforests normally immune to fire in Amazonia and Mesoamerica burned stubbornly. Winds brought the smoke from tens of thousands of fires—the largest complex on record for Mexico—in a great gyre through the southern United States. Then Florida erupted, with lightning-kindled fires in every county, and the Earth's greatest fire power, a country capable of spending a billion dollars fighting fires in a single season, was forced to evacuate 100,000 people before the taunting flames.

The fires were telegenic, they were timely. Burning Borneo, smoked-in Singapore, ravaged Russia—all seemingly became nature's metaphor for the collapse of Asia's emerging economies. The suggestion circulated further that here was a signature of global warming. An unstable climate was arcing into fire. The endless burnings were the pilot flames of an environmental apocalypse. Flaming Florida argued further that technological fixes were few and far between, that a bull market in American stocks could not halt the inexorable decline of nature's exhausted economy. The fires could not be bought off or beaten off. The future was fire.

But staring into the flames, however hypnotic, missed half the story. The Earth had known greater fire complexes, even recently. The 1982–83 fires in Indonesia were, in fact, larger than those of 1997–98. The Siberian fires of 1987 dwarfed those of the Far East by a factor of five or more. Amazonia had burned more seriously in 1988. Western Canada had endured more massive outbreaks in 1981, 1989, and 1994. The 3 percent of Florida's protected lands that burned paled beside the 105 percent (!) reportedly burned at the beginning of the century. Wildfire, however, was not the core concern. The flame-mesmerized media missed the fact that ENSO's climatic shuffle meant that areas that normally burned did not. That was emblematic of the great expanses of Earth that no longer accepted routine fire. For most of the planet, 1998 was once again "The Year the Earth Hardly Burned."

What the millennium displayed was a colossal maldistribution of combustion—too much of the wrong fire, too little of the right. In general, the developing world had too much wildfire, the developed world too little controlled burning. El Niño's climatic rhythms had an echo: there were places of fire drought as well as fire deluge. But the eruption of wildfire followed climatic rhythms, while the erosion of routine burning obeyed a deeper driver. Behind it, like the ponderous changes of climate that swung into and out of ice ages, hummed the dynamo of industrial fire. Probably the Earth was experiencing as much or more combustion than it ever had because there was more fuel to burn. Paradoxically, it knew less free-burning fire than it had since the last millennium, perhaps since the retreat of the Pleistocene ice. That rending of combustion from flame explained a lot about why the Earth was burning as it was.

As the World Burns: What Is and Isn't Burning, and Where

Places with Too Much Fire

Places with too much fire were, by and large, places with lots of fresh fuel. Of course, timing mattered, and that meant weather. Combustibles could pile up from logging or land clearing, but if they remained wet, they didn't burn. Similarly, a severe drought could, by itself, stock a landscape with fuel from stress-dropped leaves and shriveled shrubs. But most of the Earth's hotspots were places in which humans, by arriving or abandoning, had allowed fuels to reach hefty amounts. They were landscapes characterized by rapid changes in land use.

New lands. The most notorious examples, because they housed exotic biotas, were tropical rainforests abruptly converted to farm and pasture. Brazil and Indonesia became the best known, not only because they contained a rich fraction of remaining rainforest but because they did so within the context of a single nation and actively promoted internal colonization for geopolitical ends. In particular, they combined clearing with schemes that sought to move people from overly dense sites to sparsely inhabited ones. Access meant roads, and roads brought loggers, farmers, and herders. The slashed biota, baked under an equatorial sun, powered fires. Yet there were plenty of other places—tropical Africa, for example—where efforts to industrialize and rapid population growth combined to crack open closed forests. Fires, once started, created conditions that led to more fires.

This is how agriculture has always expanded into new land. Fire catalyzes, fire removes, fire transmutes. What differs now is the global attention the process receives. There are television cameras to broadcast the scene, environmentalist groups to protest, and global values—a kind of green equivalent of human rights—to argue for the preservation of biodiversity. There are political parties to speak also for native peoples, because the lands into which the bulldozers and newcomers moved are often not truly uninhabited. There are meteorologists to track the smog palls that settled onto Jakarta and Singapore and climatologists to measure greenhouse gases that, once freed from their biotic fetters, drifted over New York and Rome. There are urban values against which to assess old rural rites. The putative primitiveness of the scene—the contrast between belching chain saw and green lianas dripping with silence—sharpens the sense of collective outrage. And, not least, there is history. Similar past events in industrialized nations evoke a dark memory of how not to remake landscapes. The scenes have become as globalized as the methane molecules they release to the atmosphere.

Old lands. Yet even as agriculture colonized new lands, it was abandoning old ones. Leaving lands alone bred fuels as surely as felling their woods. Abandoned agricultural fields overgrew with scrub, and as they did they became a habitat for wildfire. In places prone to fire, the garden-gone-to-weed was creating global hotspots as vigorous as those subject to ax and bulldozer.

There were two variants, one obvious, the other less so. The self-apparent version was the Mediterranean region, at least that half not overpopulated with people and flocks to the point that every scrap of biomass ended up in hearth or gullet. The once close-tended plots blossomed into rough combustibles, and the formerly close-guarded fire gorged on the litter. The domesticated fire went feral. Something similar happened in the exurbs of developed nations, as abandoned fields and pastures sprouted to houses, a lusty stockpile of fuel that wildfire soon discovered. This "intermix" fire, rampant in Australia and America, was the fraternal twin of the fire scorching the mountains of Greece and Provence.

The sharper contrast dwelled in places deliberately created as wildlands. No gradual withdrawal here, no waning of field and flock, no waxing of scrub and houses. These lands were reserved suddenly, by calculated choice, from human habitation, and where necessary this was backed by political force. Whereas humans were trucked into Borneo and Rondônia to establish permanent settlements, they were excluded from Ngorongoro Crater, the Teton National Forest, and the Barguzinskii *zapovednik* except as leave-no-trace transients. But because there were no leaping flames or towering smoke, kindled by human ambition, to track the fire scene and stir public passions, the crisis lay dormant.

Although it was possible to stop the flow of people, it was not possible to stop the growth of plants. Over time, sites, even monstrously degraded ones, recovered. As vegetation reclaimed these places, they nourished a trophic hierarchy of fuels, a pyrodiversity of combustibles. The banning of fire from such sites was not always an option. In time, many would burn. Too often they burned too fiercely and at the wrong time.

Lost control. Elsewhere, fires broke out in places that lost, or chose to withdraw, their ability to control unwanted fires. Social upheavals could cripple those institutions created to suppress fire, as governments collapsed, political purposes changed, and economies imploded. Particularly on public wildlands, anything that affected state institutions could affect their ability to fight wildfire. That might be as mild as a change of policy, or as profound as a change of government or the wrenching move from a closed to a market economy.

That Canada experienced sharp peaks in burned area beginning around 1980 largely as a result of shifts in fire policy. For decades the provinces had an implied boundary beyond which they had left fires to burn. In the late 1970s, however, they began to attack fires north of that line. For a while the enterprise worked. Then came doubts about its costs, both economic and ecological, and suppression forces withdrew behind their old border. Aerial observation continued, however, so the fires, often vast but no longer fought, entered the national ledger of area burned. Canada's surprising surge of fires was in fact an outcome of accounting and policy. The boreal forest had always burned, sometimes hugely. Now there were people to observe it.*

The outbreaks in the former Soviet Union and Mongolia were more likely real in that they represented fires that might not have started so abundantly or grown so big had the previous regime remained in place. The story is complicated because the statistical record is unreliable. The old regimes had suppressed fire reports as much as fires. Also by redefining what lands were formally "under protection," it was possible to add or delete fires from the official count. Still, the large-year spikes are most likely at least partly a result of the political upheaval that led to the collapse of the Soviet Union and pulled Mongolia out from under the tutelage of the USSR and communism. A paramilitary force, funded by a command economy for political ends, forest fire protection had suited the Soviet model nicely.

Before 1991, the Soviet Union had the largest aerial firefighting force in the world, most of it posted in Siberia and the Far East. The same system, adapted, fought fires in Mongolia. Both have been caught in a fast spiral of decline, worsened by mountains of logging slash strewn as a result of their countries' wild entry into a global market. More sparsely settled, Mongolia has felt the shock more keenly. Rashes of fires broke out from new causes, such as the scrounging in early spring for elk antlers (sold to Europeans and Chinese). Other fires lingered malignantly on the land, perhaps because collectivization broke down and with it the organization it had imposed, however harshly, on rural life. Mongolian wildfires, especially, have spun out of control. In 1996 and 1997, burned area increased 18 to 20 times the annual average. Together the two years' fire scorched more forest land than had been harvested in the last 65 years.†

*For the Canadian statistics, see Brian J. Stocks, "The Extent and Impact of Forest Fires in Northern Circumpolar Countries," in Joel L. Levine, ed., *Global Biomass Burning* (Cambridge: MIT Press, 1991), pp. 197–202.

†See E. N. Valendik et al., "Fire in Forest Ecosystems in Mongolia," *International Forest Fire News* 19 (September 1998): 58–63; and James R. Wingard and N. Erdenesaikhan, "The German-Mongolian Technical Cooperation GTZ Integrated Fire Management Project, Khan Khentii Protected Area, Mongolia," *International Forest Fire News* 19 (September 1998): 64–66.

Places with Too Little Fire

Oddly, given the tenor of media coverage, most of the planet suffered degrees of fire famine. The Earth's fire excesses had their complement in fire deficits. If those missing fires were less visible, they were no less significant ecologically. On the scale of landscapes, Third Fire could not substitute for all of fire's ecological effects. Those absent flames were fire's quiet crisis in the developed world.

The origin of the fires didn't matter, save to environmental *philosophes*. Removing natural fire, suppressing aboriginal fire, squelching agricultural fire—all disturbed the fire regime, and their reduction had consequences. The biota overgrew, it restructured, and it replaced some species with others. But only rarely did fire truly disappear, for the simple reason that few landscapes tolerate a fire vacuum. Some fire—natural, anthropogenic, industrial—will fill the void. Thus places often suffered first because they lacked the kind of fire to which they had grown accustomed, and second because fire, when it eventually entered, too frequently burned with a ferocity that gutted rather than renewed the scene.

As the issue evolved, two landscapes dominated, one each from the most industrialized continents, Europe and North America. Long-settled Europe struggled to preserve vestiges of its cultural landscapes, nearly all of them agricultural and virtually all reliant on an enabling fire. Mowing, grazing, cutting, manuring, browsing—none completely restored the old scenes. The reason was that those sites had also burned. Fire, rightly intertwined with the other practices, had become a necessary though not by itself sufficient cause of land renewal. This was not at first obvious.*

Gradually, however, as industrialization replaced traditional agriculture, and fossil fallow replaced living fallow, open fire—long distrusted by Europe's intellectuals and officials—sank into vestigial embers. Landowners found it increasingly difficult to sustain moors, upland meadows,

*For Europe, the story is unfolding in the pages of *International Forest Fire News* and various regional conferences. An introduction to the issues, particularly good because it deals with central Europe, is *Feuereinsatz im Naturschutz*, NNA Berichte (Schneverdingen, Niedersachsen: Alfred Toepfer Akademie für Naturschutz, 1997).

For America the best source is the Tall Timbers Fire Ecology Conference proceedings, beginning in 1962. For fire in wilderness, two published symposia serve as a good guide to the arguments: James E. Lotan et al., tech. coords., *Proceedings: Symposium and Workshop on Wilderness Fire*, Gen. Tech. Report INT-182 (U.S. Forest Service, 1985), and James K. Brown et al., tech. coords., *Proceedings: Symposium on Fire in Wilderness and Park Management*, Gen. Tech. Report INT-GTR-320 (U.S. Forest Service, 1995).

farmed-and-grazed woodlands, and mixes of flora and fauna that had depended on a landscape mosaic of patches of varying ages. Many of the desired scenes had derived from swiddening and long fallow, or from burning and close browsing. At some point cultivated fire had to return.

America appealed instead to wilderness, to a nature that humans did not occupy nor their arts shape. However glorious in principle, such landscapes often proved illusory in practice. The preserved places had at least partly resulted from past human use, most often linked to burning. Still, the belief dawned that natural fire had a valid right to flourish in such settings. So when officials sought to "restore" fire, they allowed natural fire but no other, except where they had no choice. The outcome was mixed, as one might expect. Too often, lightning failed to kindle enough fires; too often, nominally "controlled" fires escaped. All too commonly the unburned biota had grown a fuel complex unlike what had existed previously, and thus one that burned very differently—either too feebly or too violently. The restoration of fire as an untrammeled ecological process failed because First Fire had not inscribed the dominant fire regime. Second Fire had.

So where Europe had difficulties imagining fire as a legitimate biological agency, America found it equally vexing to imagine humans as a legitimate ecological agent. Europe thus groped to put fire back into the system; America to insert and integrate people. Worse, by the time the two traditions began to converge in their thinking, industrial combustion had so progressed that any attempt to reinstate fire proved exceedingly thorny. Urbanites didn't like black landscapes, didn't want smoke, didn't approve of routine burning. They could not understand fire from personal experience except as urban disaster. They were unwilling to tolerate mistakes and escaped fires. Many disliked any hint of human meddling in nature reserves. And they were properly wary about the huge costs, social and economic, associated with a program of wholesale fire restoration. At least by the turn of the millennium, they did not agree with the argument that some kind of fire would occur and that controlled fire was better than wild.

Places with Mixed Fires

Those were the polar extremes: too much and too little, lightly inhabited lands suddenly opened to fire and newly uninhabited lands abruptly closed to it. But among the spectrum of problems that lay between them was one in which the two extremes closed as industrialized societies rammed cities and wildlands together. Bureaucrats labeled it the wildland/urban interface. Others have called it more simply a zone of intermixed fire.

Behind the mingling of fires lies the mixing of fuels. The scrambling of houses and woods is nothing new. Farmers have long colonized by fracturing woods and erecting buildings. But there the excess fuels had resulted from clearing; here it sprang from new growth. Once-rural landscapes now sprout houses, an exurban sprawl, while urban areas overgrow with trees and brush. Much of the problem, that is, derives not from what people do but from what they elect not to do. They refuse to cut back the scrub, and often promote it. The upshot is an ecological omelette of fuels. And while fire is rare—controlled burning no longer a common practice of land use—more often than not flame eventually comes, and then with pent-up violence.

This peculiar environmental recipe has appeared throughout the industrialized world. Some places—temperate Europe, for example, or northeastern America—escape regular outbreaks because climatic conditions prevent routine burning. Fire exists because people choose to put it there, and as urban values extend prohibitions against wood-burning fireplaces, the burning of fallen leaves, and any open flame, there are fewer opportunities for fire to escape and gorge on the amassed fuels. But in the fynbos of South Africa, Australia's urban bush, the exurban maquis of the northern Mediterranean, and everywhere in the United States outside the fire-dead Northeast, intermix fires have burst forth with increasing regularity. Their range is impressive: the Oakland fire of 1991, the Malibu fires of 1993, the Florida fires of 1998, fires around and into the suburbs of Spokane, Reno, Missoula, Houghton, Payson.

Technical solutions are possible, and widely known. Eliminate wood-shingle roofs. Clear vegetation from around structures. Design roads for easy entry and exit. Urge insurance companies to apply pressure to enforce codes. Install a firefighting force. Create, in brief, a suitable system of building codes and fire practices to keep the landscape from erupting into deadly flame. The problem, however, is not one of technology or knowledge—the means to cure the disease are at hand—but of values. For although urbanites are recolonizing once-rural landscapes, they are not living off a rural economy. These are people who want woody jungles for privacy and naturalness, who are fleeing city taxes and bureaucracies, who are transients and reside only seasonally, who value solitude from the crush of modern life; people for whom the shaggy scene, often overripe with combustibles, is precisely what they seek. They believe that open fire and drifting smoke are unnecessary and dangerous, that the fire codes and values of urban life can apply to exurban existence as well.

Behind such beliefs stands the accrued experience of industrialization. These are populations that have grown up in cities and seen firsthand the power of modern devices to replace old practices. They might value a ceremonial fireplace, but they heat their houses with electricity or natural gas. They would clean up debris with chain saws or chippers, or haul pine needles to a landfill in a truck; they would not run free-burning fire over the scene. The landscape is thus metastable: it is a land changing to something else, most likely destined to burn anew or eventually evolve into less combustible forms. In this the landscape mirrors industrialization, for the mechanical hand of Third Fire has created both.

A Planet with Too Much Combustion and Too Little Fire

More than open fires misplaced, missing, or mixed, industrial fire has emerged at the onset of the third millennium as the driver of planetary combustion. More and more, fire appears less as something that results from climate and increasingly as something that shapes climate. No longer does combustion seem contained within ancient ecological equilibriums, however those might be defined. It grows exponentially with its endless, exhumed fuels. Critics claim that even wildfires—those that gorge on cleared rainforest, those that sweep through surging scrub—are themselves ecological aftertremors of the industrial quake that has shaken the Earth.

Third Fire's ashes and fumes are overfilling the biological sinks allotted for burning. Greenhouse gases are stuffing a thin atmosphere, not only with carbon dioxide but with more exotic (and potent) gases like bromine released by wildfire and like methane pumped out by organisms that thrive after the burn. But there are other, worrisome outcomes as well: acid rain, poor visibility, tropospheric ozone, threats to human health from combustion-generated particulates. All of this toxic cocktail free-burning fire also yields, but with seasonal timing and with an evolutionary history of adaptations and in quantities that biotas and the planetary atmosphere and oceans can absorb. Their shared chemistry—their competition for combustion sinks—has put fossil-fuel industrial fire into direct rivalry with biomass-fueled open fire. One way to reduce the impacts of industrial fire, critics argue, is to reduce other forms of burning.

The Kyoto Protocol commits its signatories to stabilize and eventually reduce greenhouse-gas emissions. The source of the gases doesn't matter. A molecule of carbon dioxide loosed by land clearing in Sumatra is no different from molecules released by charcoal cooking fires in India, savanna burning in Zambia, or controlled burning for Karner blue butterfly habitat in New York. In a system of carbon credits, one can trade one form of burning for another. In the United States, for example, a proposed expansion of controlled fire for ecological benefits would require a reduction in industrial emissions. The competition for combustion—hidden from most people by the machinery of modern industry—must surface as the value-laden choice it has always been.

An exponential growth in industrial fire cannot continue indefinitely. The deep question is how growth might slow, halt, or reverse itself. Is there a combustion equivalent to the demographic transition that has historically characterized societies as they industrialize? The earliest population changes typically combine high birth rates with plunging death rates, which makes the overall growth explosive. Something similar might well occur with fire. A pyric transition might show an outburst of combustion—an exorbitant use of fossil fuels that gradually replaces traditional burning. Over a period of decades, a more sustainable fire ecology would emerge. That is the optimistic scenario.

There is evidence to support it. There are places and processes where industrial fire can substitute for open flame, and there are huge improvements possible in fossil-fuel combustion. More cars, for example, do not automatically mean more carbon dioxide. In technologically advanced societies, a "decarbonization" of energy is under way such that greater energy (and fewer emissions) result from more efficient combustion of carbon-based fuels. The commercial development of fuel cells to power vehicles could prompt the combustion load of the planet to drop as quickly as the arrival of the internal combustion engine caused it to shoot up. Industrial fire threatens because it is undiluted—too singular, too recent, too disaggregated from the larger ecology of Earth. That can change.

So can the argument that this fire, like others, ought to be suppressed, that the absence of fire is intrinsically better than its presence. Contemporary attempts to regulate burning resemble strategies of nature preservation that seek to protect a place by stripping away the human presence. Yet it is not possible to reserve the atmosphere, or to preserve climate; they will change, with or without human action. Rather, it may be that future generations will seek to wield Third Fire to reshape climate as they have Second Fire in the past to reshape biotas. After all, most of the climates in which humanity has evolved have been unfavorable to human life; people relied on fire to make them livable. So, in coming centuries, we may seek to use industrial fire to render climate more agreeable, to stall the onset of new ice ages, to dampen the swings of drought and deluge. Controlled burning may extend its range to the sky. Tweaking climate may combine with engineering genes to define the macro- and micro-economics of nature's future economy.

Still the Keeper of the Flame

How Fire's Ecology Has Changed

Ever since humans first seized fire, we have used it to our own ends, remade more and more of the biosphere, and fiercely guarded our monopoly. The relationship is odd, unlike anything else on the planet—more than the bonding of an artisan with a favored tool, more than the affinity between a herder and his flock, more than an alliance of convenience. Humanity and fire have blended into an almost biological symbiosis. Nearly everywhere fire has assumed a human face and become humanity's pyric double. Since the first tread of *Homo sapiens,* fire ecology has meant human ecology.

Yet the differences between the human and the natural ecology of fire, while distinct, have grown in force with the passing millennia. Spark became both steadfast and variable. Human firebrands shifted fire's appearance within seasons, and so seized the initiative from wildfire. But Second Fire also became constant, burning year by year without the erratic slapdash of lightning fire. The existing biota adapted, sometimes hugely (as in Australia), often subtly, with trees giving way to prairies or perennial grasses to annuals. Pare people from the land and those fire regimes unravel. Explaining them without appeal to the firestick is gibberish.

Tough limits remained. Mostly, people could only work with what nature presented to them by way of weather and fuels. They could rarely bring fire where nature would not allow it. They had exploited a fire void, filling up blanks left by nature with flame. Under suitable conditions, humans could push out the frontier of Second Fire landscapes by means both brutal and delicate. But if conditions turned sour, they could not hold the flames against rain and blurred seasons. The frontier would roll back. Both happened, and both could happen again and again to a given site.

The flame's keepers knew full well both their power and its limits. The possession of fire made them unique, distinct among creatures, yet their fire power itself flowed from nature, which might inscrutably give and withhold. Their fire starters were stone, wood, bone. Their myths often told how fire leaped out of wood or flint when freed from its bondage by people. So, it seemed, had humans unfettered flame from nature's fickle thrall and then held it, as best they could, as their own.

Their grasp tightened as they gained fuller control over combustibles. They could, within bounds, make and break biomass to fashion fuel, and to that extent even defy climate. They could insert fire where it could not, under nature's sole discretion, prosper. They could transform whole landscapes into an immense, biotic hearth. So humans had, even under aboriginal conditions, wielded some control over what their torches could combust. But what nature regrew would burn only if rain and wind permitted it. Often they did not.

Agriculture proved itself a hardy traveler. It needed for its toolkit something with which to slash and something with which to cast sparks. Its ecological core was the capacity to create stuff to burn and then kindle those combustibles. The combinations of plants, animals, ax, plow, people, and fire were many. Cultivation placed fire ecology even more strenuously into human hands. How fire behaved on the planet related to the will and whim of human life, to a widening gamut of politics, trade, scholarship, legal conceptions of landownership, none of which had influenced First Fire.

Industrial fire drew combustion closer to culture. This fire, unlike others, relied little on what nature deigned to grant it. Its fire starters were a second-order technology like heated wires and electrical arcs, not natural objects. It burned within enclosed—metal and ceramic—chambers. It combusted biomass drafted from the geologic rather than the biologic realm. While humans have long held a species monopoly over free-burning fire, we never had exclusive rights to fire itself. If we leave, biomass still burns (or not). Only the regime changes.

But Third Fire cannot thrive without its human tenders. It would expire, instantly. Equally, it can burn with utter disdain for weather and whatever fuels the biota may or may not offer. It can burn as easily perched on granite or amid dripping rainforests or over storm-tossed seas. With Third Fire, we become more than the movers of ecological levers and assume the mantle of designers of novel ecosystems that cannot exist without us. More and more, the defining flow of energy through the biosphere is the flow of industrial combustion. More than ever, the mechanics of fire ecology are incomprehensible without including the mechanics of human society. What we know (or don't know, or think we know, wrongly) matters as much as the moisture content of fuels. How we move knowledge through institutions affects fire's ecology as fully as the turning of the seasons. The flow of information is as vital as the flow of nitrogen or sulfur. The structure of institutions has molded biotas as surely as mountains and rivers and the rhythm of the seasons. Scientific periodicals, professional journals, books, popular magazines, television—all have packaged and shunted the information upon which society decides how it proposes to manage fire.

A fire in South Africa can influence fire programs in Australia. The Yellowstone fires of 1988 shut down prescribed natural fire programs across the country, and gave pause to fire strategists around the world. Norman Maclean's 1992 meditation on the Mann Gulch fire of 1949, *Young Men and Fire,* followed by the mass-fatality South Canyon fire in 1994, stunned America's national fire agencies and spurred them to rethink policy on America's public lands. All this is as significant as the tidal flows of El Niño for deciding where and how fire would sit on the landscape. And since the world widely regarded the United States as a leader, if not a model, for handling Third Fire landscapes, those decisions reached far beyond America's shores.

The Long Burn

The geography of Earthly fire remains today neither exclusively natural nor exclusively human. We have not put fire in significant ways into the Sahara, save through the flaring off of natural gas. Nor have we abolished fire from the Siberian taiga. But the geography of fire looks the way it does because of what we have done and not done. That power did not originate with industrial fire. We acquired it as part of our heritage as a species. While Third Fire has prompted a change in kind, not just one of degree, the reality remains that humans have created fire's contemporary geography.

Clearly there have been epochs in which fuels have exceeded fires, in which there has been more biomass than burning. And there are times—the present age, for example—when fire combusts more than what the biosphere grows. Unfortunately, the long-term course of fire history is unknown and will probably be understood only obliquely—as charcoal buried in sediments, as gases lodged in the atmosphere, as shifting climates.

The historical contours of *Homo sapiens* as a planetary fire force, however, are better fathomed. Overall, the fire load of the planet has increased; by how much is difficult to say. In many areas, human agency has meant a change in regime, not in the absolute presence or absence of fire. Only rarely, and then very recently, have humans removed fire from any significant realm. Almost always that expulsion involves competition with, or replacement by, industrial combustion. Probably the Earth's fire load has increased over the last century, at least as measured by the flow of combustion. Ultimately even this source must shrink. Anthropogenic fire will again have to restrict itself to the cycles of what can be grown. Humanity will have to transcend Third Fire technology, as it did Second Fire, to fashion other sources of power than controlled combustion. But that prospect lies centuries in the future. It may not arrive by the end of the third millennium.

Fire has meant many things to us, and we to fire. Yet throughout the span of centuries and constantly amid all our shifting roles—suppressor of lightning fire, promoter of anthropogenic fire, stoker of industrial fire—we have remained the keeper of the planetary flame. Viewed over geologic time, our presence may appear fleeting, but measured by its ecological effects, we have had the impact of a slow collision with an asteroid, throwing embers to all sides, overturning continents, altering climates, wiping out and restoring biotas. Such is the power of fire. And whether or not it was a power we sought, much less deserved, it was a power we gained, and one we have never renounced. The seizure of fire was our most daring, our most profound gamble. It made us the biotic creatures we are.

Our prolonged crash into the biosphere has been, above all, a long burn. Beyond the next epoch of geologic time, well after this species has expired and another must examine its record, we may come to be seen as we have so often seen ourselves, as a flame—destroying, renewing, transmuting. The Earth's greatest epoch of fire will most likely coincide with our own. Unquenchable fires will have marked our passage. Charcoal will track our progress through history. The flame—tended, suppressed, abandoned—will speak uniquely to our identity as creatures of the Earth.

As it should.

SELECTED SOURCES AND FURTHER READING

Chapter 1. Fire and Earth

Bond, William J., and Brian W. van Wilgen. *Fire and Plants.* London: Chapman and Hall, 1996. Excellent survey.

Cheney, N. P., and A. M. Gill, eds. *Conference on Bushfire Modelling and Fire Danger Rating Systems: Proceedings.* Canberra, Australia: CSIRO, 1991.

Clark, James S., Helene Cachier, Johann G. Goldammer, Brian Stocks, eds. *Sediment Records of Biomass Burning and Global Change.* NATO ASI Series I, vol. 51. Berlin: Springer-Verlag, 1997. Best summary of paleofire research.

Cloud, Preston. *Oasis in Space: Earth History from the Beginning.* New York: Norton, 1988. Good on early Earth, particularly the evolution of an oxygenated atmosphere.

Crutzen, P. J., and J. G. Goldammer, eds. *Fire in the Environment: The Ecological, Atmospheric, and Climatic Importance of Vegetation Fires.* New York: Wiley, 1993.

DeBano, Leonard F., Daniel G. Neary, and Peter F. Ffolliott. *Fire's Effects on Ecosystems.* New York: Wiley, 1998.

Drysdale, Dougal. *An Introduction to Fire Dynamics.* New York: Wiley, 1985. General combustion behavior, moderately technical.

Dudley, Robert. "Atmospheric Oxygen, Giant Paleozoic Insects and the Evolution of Aerial Locomotor Performance." *Journal of Experimental Biology* 201 (1998): 1043–1050.

Komarek, E. V. "The Natural History of Lightning." Pages 139–184 in *Proceedings, Third Annual Tall Timbers Fire Ecology Conference.* Tallahassee: Tall Timbers Research Station, 1964.

Lovelock, James. *The Ages of Gaia: A Biography of Our Living Earth.* New York: Norton, 1988; Bantam Books, 1990.

Pyne, Stephen J., Patricia L. Andrews, and Richard D. Laven. *Introduction to Wildland Fire.* 2d ed. New York: Wiley, 1996.

Rossotti, Hazel. *Fire.* Oxford: Oxford University Press, 1993. Popular survey of fire in many of its forms, including pyrotechnical.

Schroeder, Mark J., and Charles C. Buck. *Fire Weather.* Agriculture Handbook 360. Washington: U.S. Forest Service, 1970. Easy entry into fire behavior; wonderfully illustrated.

Tall Timbers Research Station. *Tall Timbers Fire Ecology Conferences, Proceedings,* 21 vol. Tallahassee: Tall Timbers Research Station, 1961–present. A rich chaotic cavalcade of fire ecology; master index is available.

Yearbook of Agriculture: Climate and Man. Washington: Government Printing Office, 1941.

Touched by Fire

Chapter 2. Frontiers of Fire: (Part I)

BioScience 39(10). "Fire Impact on Yellowstone" (November 1989).

Clark, Robin. "Bushfires and Vegetation before European Settlement." Pages 61–74 in Peter Stanbury, ed., *Bushfires: Their Effect on Australian Life and Landscape.* Sydney: Macleay Museum, 1981.

Colinvaux, Paul. "The History of the Forests on the Isthmus from the Ice Age to the Present." Pages 123–136 in Anthony G. Coates, ed., *Central America: A Natural and Cultural History.* New Haven: Yale University Press, 1997.

Despain, D., et al. *A Bibliography and Directory of the Yellowstone Fires of 1988* (n.d.).

Goudsblom, Johan. *Fire and Civilization.* London: Allen Lane; New York: Penguin Books, 1992. Good treatment of species monopoly question.

Martin, Paul S., and Richard G. Klein, eds. *Quaternary Extinctions: A Prehistoric Revolution.* Tucson: University of Arizona Press, 1984. Stronger on hunting than burning, but an indispensable introduction to the data and range of explanations.

Parsons, David, and Jan van Wagtendonk. "Fire Research in the Sierra Nevada." Pages 35–46 in William L. Halvorson and Gary E. Davis, eds., *Science and Ecosystem Management in the National Parks.* Tucson: University of Arizona Press, 1996.

Trollope, W. S. W., et al. "A Structured vs. a Wilderness Approach to Burning in the Kruger National Park in South Africa." *Fifth International Rangeland Congress 1995* Denver: Society for Range Management, 1995, pp. 574-575.

Van Wilgen, B. W., H. C. Biggs, and A. L. F. Potgieter. "Fire Management and Research in the Kruger National Park, with Suggestions on the Detection of Thresholds of Potential Concern." *Koedoe* 41(1) (1998): 69–87.

Chapter 3. Aboriginal Fire

Anderson, M. K. "Prehistoric Anthropogenic Wildland Burning by Hunter-Gatherer Societies in the Temperate Regions: A New Source, Sink, or Neutral to the Global Carbon Budget?" *Chemosphere* 29(5) (1994): 913–934.

Barrett, Stephen. "Relationship of Indian-Caused Fires to the Ecology of Western Montana Forests." M.S. thesis, University of Montana, 1980.

Boyd, Robert, ed. *Indians, Fire, and the Land in the Pacific Northwest.* Corvallis: Oregon State University Press, 1999.

Gabriel, Herman W., and Gerald F. Tande. "A Regional Approach to Fire History in Alaska." BLM-Alaska Technical Report 9 (September 1983). Anchorage: U.S. Department of the Interior, Bureau of Land Management.

Hallam, Sylvia J. *Fire and Hearth: A Study of Aboriginal Usage and European Usurpation in South-western Australia.* Canberra: Australian Institute of Aboriginal Studies, 1979.

Jones, Rhys. "Fire Stick Farming." *Australian Natural History* 16 (1969): 224–228.

Lewis, H. T. *A Time for Burning: Traditional Indian Uses of Fire in the Western Canadian Boreal Forest.* Edmonton: Boreal Institute for Northern Studies, University of Alberta, 1982.

Lewis, H. T., and T. M. Ferguson. "Yards, Corridors, and Mosaics: How to Burn a Boreal Forest." *Human Ecology* 16 (1988): 57–77.

Ward, David, and Rick Sneeuwjagt. "Believing the Balga." *LANDSCOPE* (Autumn 1999), reprint.

White, Clifford A., and Ian R. Pengelly. "Fire as a Natural Process and a Management Tool: The Banff National Park Experience." Pages 54–69 in Dawn Dickinson et al., eds., *Proceedings of the Cypress Hills Forest Management Workshop.* Medicine Hat, Alberta: Society of Grasslands Naturalists, 1992.

Rites of Fire

Frazer, Sir James. *Balder the Beautiful: The Fire-Festivals of Europe and the Doctrine of the External Soul.* 2 vols. New York: Macmillan and Co., 1923.

———. "The Prytaneum, the Temple of Vesta, the Vestal Fire, Perpetual Fires." *Journal of Philology* 14 (1885): 145–172.

Chapter 4. Agricultural Fire
Chapter 5. Frontiers of Fire (Part 2)

Batchelder, Robert B., and Howard F. Hirt. *Fire in Tropical Forests and Grasslands.* Tech. Report 67-41-ES. Natick, Mass.: U.S. Army Natick Laboratories, 1966.

Bradshaw, R. H. W., K. Tolonen, and M. Tolonen. "Holocene Records of Fire from the Boreal and Temperate Zones of Europe." Pages 347–365 in James S. Clark et al., *Sediment Records of Biomass Burning and Global Change.* NATO ASI Series I, vol. 51. Berlin: Springer-Verlag, 1997.

Braudel, Fernand. *The Mediterranean and the Mediterranean World in the Age of Philip II.* 2 vols. Translated by Sian Reynolds. New York: Harper and Row, 1972. See vol. 1, pp. 85–102, for a survey of transhumance.
Bringeus, Nils-Arvid. *Brännodling: En historik-ethnologisk undersökning.* Skrifter från folklisvsarkivet I Lund. Utgivne genom sällkskpet folkkultur, 6. Lund, 1963.
Conklin, H. C. "The Study of Shifting Cultivation." *Current Anthropology* 1 (1961): 27–61.
Evans, E. Estyn. "Transhumance in Europe." *Geography* 25 (1940): 172–180.
Heikinheimo, Olli. *Kaskeamisen vaikutus Suomen metsiin.* Helsinki, 1915.
Iversen, Johannes. *The Development of Denmark's Nature since the Last Glacial.* Translated by Michael Robson. DGU V. Series, 7-C. Copenhagen, 1973.
Lewis, H. T. "The Role of Fire in the Domestication of Plants and Animals in Southwest Asia: A Hypothesis." *Man* 7 (1972): 195–222.
Sigaut, François. *L'Agriculture et le feu: Role et place du feu dans les techniques de préparation du champ de l'ancienne agriculture européenne.* Paris: Mouton and Co., 1975.
Soininen, Arvro M. "Burn-Beating as the Technical Basis of Colonisation in Finland in the 16th and 17th Centuries." *Scandinavian Economic History Review* 7 (1959): 150–166.
Steensberg, Axel. *Fire Clearance Husbandry: Traditional Techniques Throughout the World.* Herning: Poul Kristensen, 1993.
Suomen Antropologi 4. Special Issue on Swidden Cultivation (1987).

Chapter 6. Urban Fire

I know of no scholarly history of urban fire through the ages, and certainly none that considers the setting from the perspective of fire ecology. The literature consists mostly of technical studies—engineering, fire dynamics—or heroic accounts of big fires and brave firefighters. The intermix fire scene is no exception, though most observers approach it from the wildland rather than the urban perspective. I consider this chapter an exploratory essay in what the field might contain.

Cowie, Leonard W. *Plague and Fire:–London, 1665–66.* East Sussex: Wayland Publishers; New York: Putnam, 1970.
Fischer, William C., and Stephen F. Arno, comps. *Protecting People and Homes from Wildfire in the Interior West: Proceedings of the Symposium and Workshop.* Gen. Tech. Report INT-251. Ogden, Utah: U.S. Forest Service, 1988. One of the first of many conferences on the subject, but complete, for all that.
Frost, L. E., and E. L. Jones. "The Fire Gap and the Greater Durability of Nineteenth-century Cities." *Planning Perspectives* 4 (1989): 333–347.

Goudsblom, Johan. *Fire and Civilization.* London: Allen Lane; New York: Penguin Books, 1992. A good introduction to European urban fire.
Jones, E. L. "The Reduction of Fire Damage in Southern England, 1650–1850." *Post-Medieval Archeology* 2 (1968): 140–149.
Lyons, John W. *Fire.* New York: Scientific American Books, 1985. Excellent summary of fires in hearths, rooms, and buildings.
Lyons, Paul Robert. *Fire in America!* Boston: National Fire Protection Association, 1976.

Chapter 7. Pyrotechnics

I know of no single or common source, since the topic is not generally recognized as a coherent field. Again, I consider the chapter an exploratory essay in the subject.

Agricola, Georgius. *De Re Metallica.* Translated by Herbert Clark Hoover and Lou Henry Hoover. New York: Dover, 1950, reprint.
Bond, Horatio, ed. *Fire and the Air War.* Boston: National Fire Protection Association, 1946.
Fisher, George J. B. *Incendiary Warfare.* London: McGraw-Hill, 1946.
Gillispie, Charles C. *A Diderot Pictorial Encyclopedia of Trades and Industry.* New York; Dover, 1987.
Lyons, John W. *Fire.* New York: Scientific American Books, 1985. Includes a number of fire appliances.
Magnus, Olaus. *Historia om de Nordiska Folken.* Mälmo: Gidlunds förlag, 1982.
Rossotti, Hazel. *Fire.* Oxford: Oxford University Press, 1993. Treats various fire devices, along with fireworks.
Smith, Cyril Stanley, and Martha Teach Gnudi, trans. and eds. *The Pirotechnia of Vannoccio Biringuccio.* Cambridge: MIT Press, 1966; New York: Dover, 1990, reprint.
Wertime, Theodore A., and Steven F. Wertime, eds. *Early Pyrotechnology: The Evolution of the First Fire-Using Industries.* Washington: Smithsonian Institution Press, 1982.

Fire in the Mind

Bachelard, Gaston. *The Psychoanalysis of Fire.* Translated by Alan C. M. Ross. Boston; Beacon Press, 1964.
Frazer, Sir James. *Myths of the Origin of Fire.* New York: Hacker Art Books, 1974, reprint.
Gergory, Joshua. *Combustion from Heracleitos to Lavoisier.* London: Edward Arnold and Co., 1934.

Chapter 8. Frontiers of Fire

To my knowledge, no systematic survey of Europe's fire expansion exists. Instead, one must consult separate national histories, the scientific literature, and histories of professions like forestry. Perhaps my Cycle books offer the best introduction, with an overview furnished in *Vestal Fire.*

Crosby, Alfred. *Ecological Imperialism.* Cambridge: Cambridge University Press, 1986.

Fernow, Bernhard. *A Brief History of Forestry in Europe, the United States, and Other Countries.* Toronto: University of Toronto Press, 1907.

Grove, Richard. *Green Imperialism: Colonial Expansion, Tropical Island Edens, and the Origins of Environmentalism, 1600–1860.* Cambridge: Cambridge University Press, 1995.

Jordan, Terry G., and Matti Kaups. *The American Backwoods Frontier: An Ethnic and Ecological Interpretation.* Baltimore: Johns Hopkins University Press, 1989.

Tucker, Richard P., and J. F. Richards, eds. *Global Deforestation and the Nineteenth-Century World Economy.* Durham, N.C.: Duke Press Policy Studies, 1983.

Turner II, B. L., et al. *The Earth as Transformed by Human Action: Global and Regional Changes in the Biosphere over the Past 300 Years.* Cambridge: Cambridge University Press, 1990.

World Conservation Monitoring Centre. *1990 United Nations List of National Parks and Protected Areas.* Gland, Switzerland: International Union for the Conservation of Nature, 1990.

Chapter 9. Industrial Fire

Crutzen, P. J., and J. G. Goldammer, eds. *Fire in the Environment.* New York: Wiley, 1993.

King, Anthony W., et al. "The Response of Atmospheric CO_2 to Changes in Land Use." Pages 326–338 in Joel Levine, ed., *Global Biomass Burning.* Cambridge: MIT Press, 1991.

Levine, Joel, ed. *Biomass Burning and Global Change.* Cambridge: MIT Press, 1996.

———. *Global Biomass Burning: Atmospheric, Climatic, and Biospheric Implications.* Cambridge: MIT Press, 1991.

Smil, Vaclav. *Energy in World History.* Boulder: Westview Press, 1994.

Smith, Kirk R. *Biofuels, Air Pollution, and Health: A Global Review.* New York: Plenum Press, 1987.

Chapter 10. The Future of Fire

See also the two volumes edited by Levine (above) for synopses of how fire looks to those seeking a global overview.

Andreae, Meinrat O. "Biomass Burning: Its History, Use, and Distribution and Its Impact on Environmental Quality and Global Climate." Pages 3–21 in Joel Levine, ed., *Global Biomass Burning.* Cambridge: MIT Press, 1991.

Biswell, Harold H. *Prescribed Burning in California Wildlands Vegetation Management.* Berkeley: University of California Press, 1989.

Bureau of the Census. *Historical Statistics of the United States: Colonial Times to 1970.* Washington: Government Printing Office, 1975.

Goldammer, J. G., ed. *Fire in the Tropical Biota.* Berlin: Springer-Verlag, 1990.

Goldammer, J. G., and J. J. Jenkins, eds. *Fire in Ecosystem Dynamics: Mediterranean and Northern Perspectives.* The Hague: SPB Academic Publishing, 1990.

Goldammer, Johann Georg, and Valentin V. Furyaev, eds. *Fire in Ecosystems of Boreal Eurasia.* Boston: Kluwer Academic Publishers, 1996.

Lim, B., and I. Renberg. "Lake Sediment Records of Fossil Fuel–Derived Carbonaceous Aerosols from Combustion." Pages 443–459 in James S. Clark et al., eds., *Sediment Records of Biomass Burning and Global Change.* Berlin: Springer-Verlag, 1997.

UN FAO, ECE, ILO. *Seminar on Forest Fire Prevention, Land Use and People.* Athens: Ministry of Agriculture, 1992.

Valendik, E. N., et al. "Fire in Forest Ecosystems in Mongolia." *International Forest Fire News* 19 (September 1998): 58–63.

Wingard, James R., and N. Erdenesaikhan. "The German-Mongolian Technical Cooperation GTZ Integrated Fire Management Project, Khan Khentii Protected Area, Mongolia." *International Forest Fire News* 19 (September 1998): 64–66.

INDEX

Boldface page numbers refer to illustrations.